Hands-On Internet of Things with Blynk

Build on the power of Blynk to configure smart devices and build exciting IoT projects

Pradeeka Seneviratne

BIRMINGHAM - MUMBAI

Hands-On Internet of Things with Blynk

Commissioning Editor: Gebin George
Acquisition Editor: Prachi Bisht
Content Development Editor: Trusha Shriyan
Technical Editor: Varsha Shivhare
Copy Editor: Safis Editing
Project Coordinator: Kinjal Bari
Proofreader: Safis Editing
Indexer: Pratik Shirodkar
Graphics: Jisha Chirayil
Production Coordinator: Shantanu Zagade

First published: May 2018

Production reference: 1220518

Published by Packt Publishing Ltd.
Livery Place
35 Livery Street
Birmingham
B3 2PB, UK.

ISBN 978-1-78899-506-1

www.packtpub.com

`mapt.io`

Mapt is an online digital library that gives you full access to over 5,000 books and videos, as well as industry leading tools to help you plan your personal development and advance your career. For more information, please visit our website.

Why subscribe?

- Spend less time learning and more time coding with practical eBooks and Videos from over 4,000 industry professionals

- Improve your learning with Skill Plans built especially for you

- Get a free eBook or video every month

- Mapt is fully searchable

- Copy and paste, print, and bookmark content

PacktPub.com

Did you know that Packt offers eBook versions of every book published, with PDF and ePub files available? You can upgrade to the eBook version at `www.PacktPub.com` and as a print book customer, you are entitled to a discount on the eBook copy. Get in touch with us at `service@packtpub.com` for more details.

At `www.PacktPub.com`, you can also read a collection of free technical articles, sign up for a range of free newsletters, and receive exclusive discounts and offers on Packt books and eBooks.

Contributors

About the author

Pradeeka Seneviratne is a software engineer with over 10 years of experience in computer programming and systems design. He is an expert in the development of Arduino- and Raspberry Pi-based embedded systems, and is currently a full-time embedded software engineer working with embedded systems and highly scalable technologies. Previously, he worked as a software engineer for several IT infrastructure and technology servicing companies. He has also authored *Beginning BBC micro:bit*, published by Apress.

About the reviewer

Munawwar Hussain Shelia is a digital artist and software engineer based in Mumbai, India. An engineering graduate (BE) in computer science, his curiosity to try his hand at the next latest technology keeps him thinking ahead on the boundaries of coding. He started his career in mobile application development with Java and Android, before moving on to full-stack JavaScript (MEAN stack) Node.js, where he put his heart and soul into building more real-time web applications. Currently, he is in the role of data science, and extracts actionable intelligence and does predictive analytics from large datasets.

Packt is searching for authors like you

If you're interested in becoming an author for Packt, please visit authors.packtpub.com and apply today. We have worked with thousands of developers and tech professionals, just like you, to help them share their insight with the global tech community. You can make a general application, apply for a specific hot topic that we are recruiting an author for, or submit your own idea.

Table of Contents

Preface

Blynk is referred to as the most user-friendly IoT platform, providing a way to build mobile applications in minutes. With Blynk's drag and drop mobile app builder, anyone can build amazing IoT applications with minimal resources and effort. Blynk supports over 400 hardware platforms and major connectivity types. The hardware could be prototyping platforms, such as Arduino and Raspberry Pi, to industrial-grade ESP8266, Intel, Sierra Wireless, Particle, and Texas Instruments offerings.

This book uses Raspberry Pi as the main hardware platform and C++ for writing code to build projects.

The first part of this book offers how to set up the development environment with Raspberry Pi, Raspbian Stretch LITE, and various software components. Then, the reader will build the first IoT application with Blynk.

The middle part of the book presents how to use and configure various widgets (control, display, and notify) with Blynk app builder to build applications.

The latter part of the book will introduce how to connect with and use built-in sensors on mobile devices such as Android and iOS. After this, the reader will learn how to set up a personal Blynk server on Raspberry Pi. Finally, the reader will learn how to build a robot vehicle that can be controlled with a Blynk app through the Blynk cloud service.

Who this book is for

This book is for those who want to build rapid IoT applications in minutes for connected products and services with only a basic understanding of electronics, Raspberry Pi, and C++.

What this book covers

Chapter 1, *Setting Up a Development Environment*, explains how to set up the development environment for Blynk with Raspberry Pi. It describes how to install Blynk libraries and some supporting software components that you can use to build Raspberry Pi-based IoT hardware. Then, you will build a control application with Blynk app builder. After that, you need to write a C++ application to connect with the Blynk cloud. Finally, you run the Blynk app to connect the Raspberry Pi to Blynk app builder through the Blynk cloud over a Wi-Fi network.

Chapter 2, *Building Your First Blynk Application*, explains how to build your first Blynk application to control an LED (or any actuator) attached to the Raspberry Pi from your smartphone or tablet. First, you will build an app with the Blynk app builder. Then, you will use digital or virtual pins to control the attached LED. After that, you will learn how to write a simple C++ application with nano text editor. Finally, you will build the application and run it to connect the Blynk app and the Raspberry Pi hardware.

Chapter 3, *Using Controller Widgets*, covers how to use controller widgets such as Slider, Step, Joystick, and zeRGBa, to control actuators. You will also learn how to use WiringPi's software PWM library, connect controller widgets with digital and virtual pins, use the split and merge mode, and parsing values coming from the controller widgets.

Chapter 4, *Using Display Widgets*, guides you on how to use display widgets, such as the Value Display widget, and Labeled Value widget to show sensor data, and the LED widget to show a button state.

Chapter 5, *Using Notification Widgets*, explains how to send notifications to the Blynk app from Raspberry Pi. You will schedule your Raspberry Pi to send notifications to your smartphone on user action. Some of the notification widgets can be integrated with third-party services, such as Twitter to send tweets from Raspberry Pi. Then, you will use the notification widget to send pop-up notifications to the smartphone or tablet. Finally, you will also learn how to send emails from Raspberry Pi using the Email widget.

Chapter 6, *Connecting with Sensors on Your Mobile Device*, guides you on how to read data from built-in sensors such as the accelerometer, light sensor, and proximity sensor on your smartphone or tablet.

Chapter 7, *Setting Up a Personal Blynk Server*, guides you on how to set up a personal Blynk server on Raspberry Pi. The Blynk personal server replaces the Blynk cloud. You can connect all your Blynk hardware to this personal server through your local network.

Chapter 8, *Controlling a Robot with Blynk*, explains how to build a robot vehicle using a two-wheeled robot chassis kit. Then, you will build an application with the Blynk app builder to control it through the Blynk cloud by connecting to a Wi-Fi network.

To get the most out of this book

You should install the Raspbian Stretch LITE operating system on Raspberry Pi. The nano text editor is used to write C++ code in the Raspberry Pi environment. PuTTY is used to make serial connections between Raspberry Pi and the computer that is running Windows.

Download the example code files

You can download the example code files for this book from your account at www.packtpub.com. If you purchased this book elsewhere, you can visit www.packtpub.com/support and register to have the files emailed directly to you.

You can download the code files by following these steps:

1. Log in or register at www.packtpub.com.
2. Select the **SUPPORT** tab.
3. Click on **Code Downloads & Errata**.
4. Enter the name of the book in the **Search** box and follow the onscreen instructions.

Once the file is downloaded, please make sure that you unzip or extract the folder using the latest version of:

- WinRAR/7-Zip for Windows
- Zipeg/iZip/UnRarX for Mac
- 7-Zip/PeaZip for Linux

The code bundle for the book is also hosted on GitHub at https://github.com/PacktPublishing/Hands-On-Internet-of-Things-with-Blynk. If there's an update to the code, it will be updated on the existing GitHub repository.

We also have other code bundles from our rich catalog of books and videos available at https://github.com/PacktPublishing/. Check them out!

Download the color images

We also provide a PDF file that has color images of the screenshots/diagrams used in this book. You can download it here: `https://www.packtpub.com/sites/default/files/downloads/HandsOnInternetofThingswithBlynk_ColorImages.pdf`.

Conventions used

There are a number of text conventions used throughout this book.

`CodeInText`: Indicates code words in text, database table names, folder names, filenames, file extensions, pathnames, dummy URLs, user input, and Twitter handles. Here is an example: "Mount the downloaded `WebStorm-10*.dmg` disk image file as another disk in your system."

A block of code is set as follows:

```
BLYNK_WRITE(V1) // zeRGBa assigned to V1
{
    // get a RED channel value
int r = param[0].asInt();
    // get a GREEN channel value
int g = param[1].asInt();
    // get a BLUE channel value
int b = param[2].asInt();
}
```

Bold: Indicates a new term, an important word, or words that you see onscreen. For example, words in menus or dialog boxes appear in the text like this. Here is an example: "Under **CONTROLLERS**, tap **zeRGBa**. A **zeRGBa** widget will add on to the canvas."

 Warnings or important notes appear like this.

 Tips and tricks appear like this.

Get in touch

Feedback from our readers is always welcome.

General feedback: Email feedback@packtpub.com and mention the book title in the subject of your message. If you have questions about any aspect of this book, please email us at questions@packtpub.com.

Errata: Although we have taken every care to ensure the accuracy of our content, mistakes do happen. If you have found a mistake in this book, we would be grateful if you would report this to us. Please visit www.packtpub.com/submit-errata, selecting your book, clicking on the Errata Submission Form link, and entering the details.

Piracy: If you come across any illegal copies of our works in any form on the Internet, we would be grateful if you would provide us with the location address or website name. Please contact us at copyright@packtpub.com with a link to the material.

If you are interested in becoming an author: If there is a topic that you have expertise in and you are interested in either writing or contributing to a book, please visit authors.packtpub.com.

Reviews

Please leave a review. Once you have read and used this book, why not leave a review on the site that you purchased it from? Potential readers can then see and use your unbiased opinion to make purchase decisions, we at Packt can understand what you think about our products, and our authors can see your feedback on their book. Thank you!

For more information about Packt, please visit packtpub.com.

Setting Up a Development Environment

Blynk is known as the most user-friendly IoT platform consisting of an app builder that can be run on iOS and Android operating systems, and a set of libraries to build amazing IoT applications in minutes with your favorite hardware platform. It allows you to quickly build interfaces by simply dragging and dropping widgets to control and monitor your hardware projects from your iOS and Android device.

In this chapter, you will learn about:

- Hardware platforms, connection types, architecture, ecosystem, and online resources
- Installing Blynk app builder on Android
- Creating a user account with the Blynk app builder to log in to Blynk Cloud
- Creating a new project with Blynk
- Preparing Raspberry Pi by adding SD card, Ethernet cable, Wi-Fi dongle, and power supply
- Writing Raspbian Stretch Lite on Raspberry Pi by flashing image to SD card with Etcher
- Configuring SSH on Raspberry Pi
- Installing prerequisite packages such as git-core and WiringPi on Raspberry Pi
- Using PuTTY to connect with Raspberry Pi with SSH protocol
- Configuring wireless connection on Raspberry Pi for Raspbian Stretch Lite
- Connecting Raspberry Pi with Blynk Cloud by running the sample C++ source file

What is Blynk?

Blynk was born as a Kickstarter project created by Pasha Baiborodin, and backed by 2,321 supporters who pledged $49,235 to help bring it to life. This is more than the finally announced stretched goal that is $26,000. Blynk is one of the successfully completed projects listed on Kickstarter (`https://www.kickstarter.com/`):

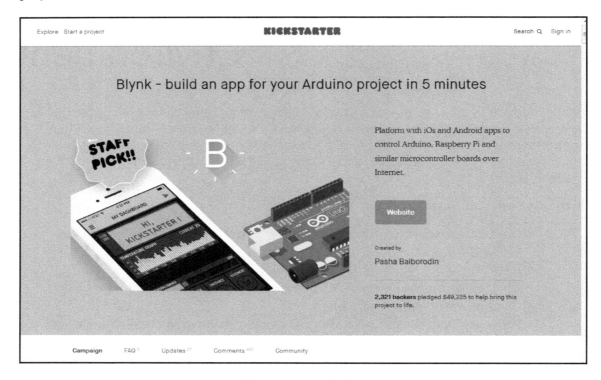

Kickstarter campaign page for Blynk

Hardware platforms

Blynk supports hardware platforms such as Arduino, Raspberry Pi, and similar microcontroller boards to build hardware for your projects. The following is a list of some microcontroller boards that can be coupled with Blynk:

- Espressif (ESP8266, ESP32, NodeMCU, WeMos D1, Adafruit HUZZAH, SparkFun Blynk Board, SparkFun ESP8266 Thing)
- Linux (C++) (Raspberry Pi, Ubuntu)

- Arduino (Arduino UNO, Arduino MKR1000, Arduino MKRZero, Arduino Yun, Arduino 101, Arduino Zero, Arduino M0, Arduino M0 Pro, Arduino Nano, Arduino Leonardo, Arduino Due, Arduino Mega 2560, Arduino Mega 1280, Arduino Mega ADK, Arduino Micro, Arduino Pro Micro, Arduino Mini, Arduino Pro Mini, Arduino Fio, Arduino Decimilia, Arduino Duemilanove, Arduino Pro, Arduino Ethernet, Arduino Leonardo ETH, Arduino Industrial 101)
- Particle (particle core, particle photon, particle electron)

You can read the list of up-to-date hardware that you can use with Blynk at `https://github.com/blynkkk/blynkkk.github.io/blob/master/SupportedHardware.md`.

Connection types

Blynk supports the following connection types to connect your microcontroller board (hardware) with the Blynk Cloud and Blynk's personal server:

- Ethernet
- Wi-Fi
- Bluetooth
- Cellular
- Serial

However, throughout this book, you will only focus on Wi-Fi and Ethernet connection types to connect with Blynk Cloud and Blynk's personal server.

Blynk architecture

The Blynk platform includes the following components:

- **Blynk app builder**: Allows to you build apps for your projects using various widgets. It is available for Android and iOS platforms.
- **Blynk server**: Responsible for all the communications between your mobile device that's running the Blynk app and the hardware. You can use the Blynk Cloud or run your private Blynk server locally. It's open source, could easily handle thousands of devices, and can even be launched on a Raspberry Pi.

- **Blynk libraries**: Enables communication with the server and processes all the incoming and outcoming commands from your Blynk app and the hardware. They are available for all the popular hardware platforms.

All the aforementioned components communicate with each other to build a fully functional IoT application that can be controlled from anywhere through a preconfigured connectivity type. You can control your hardware from the Blynk app running on your mobile device through the Blynk Cloud or Blynk's personal server. It works the same in the opposite direction by sending rows of processed data from hardware to your Blynk app.

Blynk ecosystem

The Blynk ecosystem consists of the following partners. They can cover anything from electronic components, to manufacturing and data plans:

- Intel IoT Solutions Alliance
- SparkFun Electronics
- Espressif
- Arduino
- Texas Instruments
- Proximus
- Deutshe Telekom
- Particle
- Samsung
- littleBits
- Hologram
- ThingSpeak.com
- Electric Imp
- Punch Through
- Codebender
- RedBearLab
- Wicked device
- TinyCircuits

Online resources

Blynk provides the following online resources through their website:

- **Getting started**: Guides you how to connect your Arduino with Blynk Cloud by writing sketch with Arduino's **Integrated Development Environment (IDE)** (http://www.blynk.cc/getting-started).
- **Documentation**: Provides a comprehensive documentation about how to work with the Blynk app builder, Blynk libraries, Blynk Cloud, and Blynk's personal server. This guide mainly focuses on the Arduino development environment (http://docs.blynk.cc/).
- **Forum**: There are thousands of people around the globe who are willing to help your Blynk projects through this conversation board. You can register with the Blynk forum by creating a new account with it, posting your questions, discussing with experts, helping others, and sharing your work with others (http://community.blynk.cc/).

The Blynk app builder

The Blynk app builder provides an easy way to build IoT apps that can be run on smartphones and tablets. It provides a predefined set of draggable and droppable modules known as widgets, and allows the users to make further configuration on widgets with an easy-to-use user interface. Using the app builder, you can build apps for your personal use or business purposes, as shown in the following screenshot.

If you want to build an app with Blynk for business purposes, visit
`http://www.blynk.io/business/` for more information:

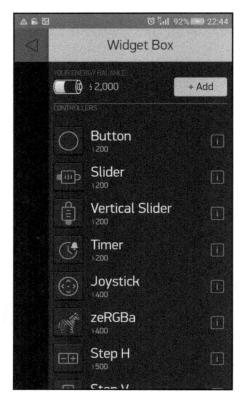

Blynk widget Box

Installing the Blynk app builder

The Blynk app builder is currently available for Android and iOS platforms. The following steps explain how to install the Blynk app builder on your Android smartphone or tablet:

1. Tap the apps icon in the lower-right corner of the home screen (consider that this is the default location).

2. Swipe left and right until you find the Play Store icon.
3. Tap the Play Store icon.
4. The first time you open the Play Store, you may be prompted to sign in with your Google credentials. If you still don't have a Google account, you can create a free account with Google.
5. Once you sign in with Google Play Store, you can search for a specific app.
6. Tap the magnifying glass in the upper-right corner, type in the name `blynk`, and tap the magnifying glass on the keyboard to execute the search. You will get a result on the screen as follows:

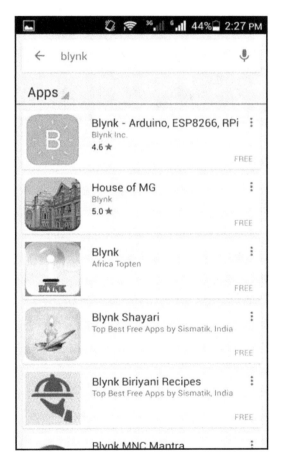

Search results for Blynk

7. Tap **INSTALL** to start the installation process. Also, you can find more information about the app by tapping the **MORE INFO** button:

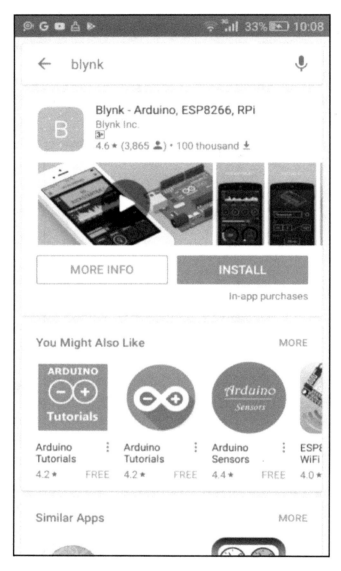

Installing Blynk app

8. Tap **ACCEPT** to allow the Blynk app to access some resources on your mobile device:

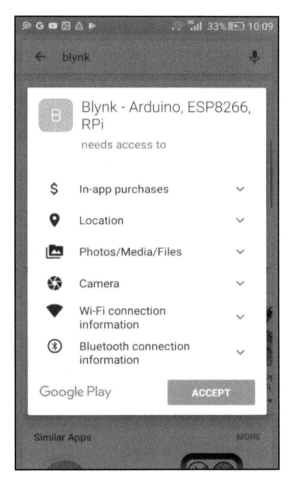

Permitting access to resources in your mobile device

9. The installation process will take some time depending on the bandwidth of your internet connection.

10. After completing the installation, you can open the Blynk app by tapping on the **OPEN** button with Google Play. This will help you to open the Blynk app builder for the first time without moving to the home screen of your smartphone or tablet:

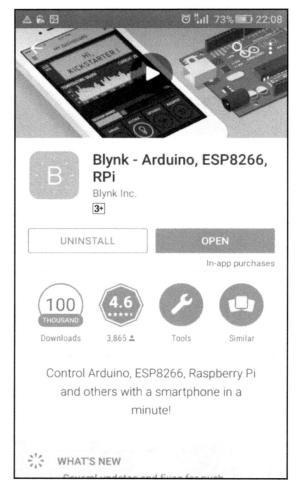

Once installed, opening Blynk app builder from Google Play

11. After installation, an application icon associated with the Blynk app builder can be found on the home screen of your smartphone or tablet:

Opening Blynk app builder from Android home screen

Creating a Blynk account

Before you start to create applications with the Blynk app builder, you should create a user account with Blynk Cloud. The following steps will guide you through how to do that:

1. Open the Blynk app builder on your mobile device by tapping on the application icon labeled **Blynk**.

2. Tap **Create New Account**.

3. Alternatively, you can log into the Blynk app builder with your Facebook account by tapping on the **Log In with Facebook** button:

Creating new account—step 1

4. On the **Create New Account** page, type your email address for the username, and a password.

5. Then, tap the **Sign Up** button:

Creating new account—step 2

6. A user account will be created, and you will automatically be logged in to the Blynk app builder (Blynk Cloud).

7. First, you will be notified on how energy works. Tap on the **Cool! Got it.** button:

Notification about how energy works

 Each widget needs energy to operate. When you delete a widget, energy is returned.

8. At any time, you can log out from the app builder by tapping the log out icon in the toolbar. When prompted, tap **LOG OUT**, or tap **CANCEL** if you want to stay logged in.

 At the time of writing, Blynk doesn't send any email message to your email account confirming the creation of the new user account.

9. When you log into the Blynk app builder again, type in your email address associated with the Blynk account and the password, then tap **Log In**.

10. You can reset your password by tapping **Reset Password** at any time.

Creating a new project

Now that you have created a new account with Blynk Cloud and logged in to your Blynk app builder running on your smartphone or tablet, you can follow the following steps to create a new project with the Blynk app builder:

1. To start building a project, tap **New Project**. This will bring you to the create **New Project** wizard:

Creating a new project

2. Type in demo in the **Project Name** textbox:

Creating new project—project name

3. Tap the **CHOOSE DEVICE** dropdown box to see the available hardware models (the default is ESP8266). Then, tap **Raspberry Pi3 B** from the **Select your hardware** list. You may find other versions of the Raspberry Pi as well. Following are the available options for Raspberry Pi versions:
   - **Raspberry Pi 2/A+/B+**
   - **Raspberry Pi 3 B**
   - **Raspberry Pi A/B (Rev2)**
   - **Raspberry Pi B (Rev 1)**

4. After choosing the hardware model, tap **OK**:

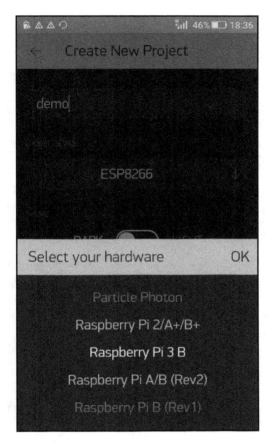

Creating new project—Selecting hardware

5. Tap on the **CONNECTION TYPE** dropdown list to see the available connection types. Then select **WiFi** from the **CONNECTION TYPE** list, followed by **OK**.

6. Tap the **Create** button to create the project workspace:

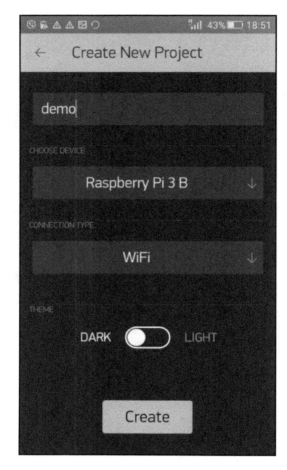

Creating new project

7. After successfully creating the project, the app builder will send an auth token to confirm your project. Every new project you create will have its own auth token. The auth token is a unique identifier that consists of alphanumeric characters, and authorizes you to access the Blynk Cloud, located at `https://www.blynk.cc/`:

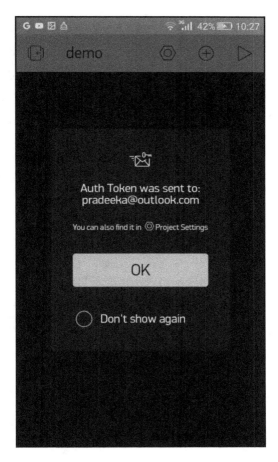

Notification about the auth token

8. At the same time, you'll have an auth token automatically sent to the email address you provided.

If you have lost your auth token, you can still find it with your project in **Project Settings** (the nut icon).

9. Tap the nut icon in the toolbar to get to the **Project Settings** page. Then, scroll down to find the **DEVICES** section. Tap **demo Raspberry Pi 3 B (WiFi)**.

10. Alternatively, you can tap **Email all** under **AUTH TOKENS** to get all the Auth tokens on your email provided for each project in your Blynk account. You can also tap **Copy all** to copy all the Auth tokens to your mobile device's clipboard:

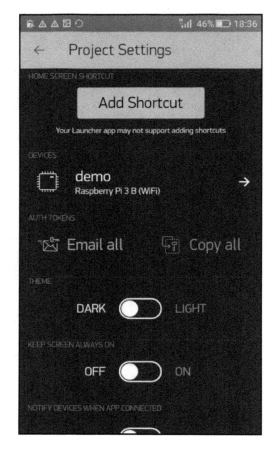

Finding Auth token—step 10

11. On the **My Devices** page, tap **demo Raspberry Pi 3 B(WiFi)** under **My Devices**:

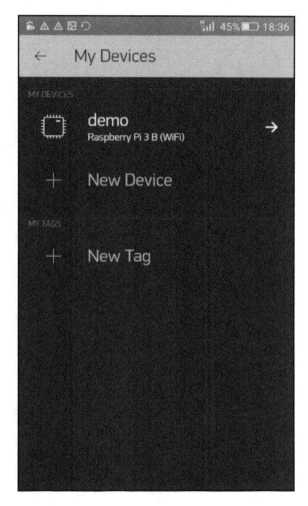

My Devices page

12. You can find the Auth token under the **AUTH TOKEN** section, but it only shows the last four alphanumeric characters. You can tap the **E-mail** button to send it to the email account again:

Finding Auth token—step 12

You will learn how to use the Auth token with your C++ code later in this chapter.

Getting parts

Blynk supports over 400 hardware platforms and major connectivity types such as Ethernet, Wi-Fi, Bluetooth, Bluetooth Low Energy, and USB; however, it is very difficult to explain how to build devices for Blynk using all the supporting hardware platforms in a single book. Therefore, this book will be using Raspberry Pi as the hardware platform to demonstrate the ability and power of Blynk.

Raspberry Pi

Raspberry Pi is a small, programmable device. It is a mixture between a very small computer and a programmable embedded board. You can connect the Raspberry Pi to the internet through Wi-Fi or Ethernet. You will be using Raspberry Pi 3, the latest model of the Raspberry Pi family, as the hardware platform; however, you can use other versions of Raspberry Pi, such as Raspberry Pi 2/A+/B+ and Raspberry Pi A/B (revision 2) to work with projects that we will be discussing throughout the book.

The sample code provided with this book is not tested with Raspberry Pi Zero W or Raspberry Pi Zero.

Here is a list of the Raspberry Pi boards available in reverse chronological order:

- Raspberry Pi 3 Model B (third-generation single-board computer)
- Raspberry Pi 2 Model B (the Raspberry Pi 2 Model B is the second-generation Raspberry Pi)
- Raspberry Pi 1 Model B+ (the model B+ is the final revision of the original Raspberry Pi)
- Raspberry Pi 1 Model A+ (the model A+ is the low-cost variant of Raspberry Pi)
- Raspberry Pi Zero W (single-board computer with wireless and Bluetooth connectivity)
- Raspberry Pi Zero (lowest-cost single-board computer)

Raspberry Pi 3 Model B, which is one of the latest models of the Raspberry Pi family, can be found here along with the other models: https://www.raspberrypi.org/products/raspberry-pi-3-model-b/.

Refer to the following screenshot:

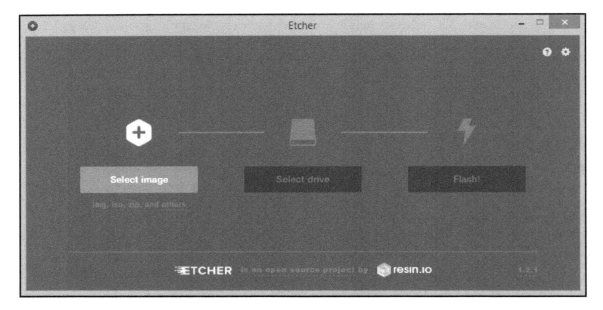

Raspberry Pi 3 Model B, a single-board computer with wireless LAN and Bluetooth connectivity

Following is the list of important specifications of Raspberry Pi Model 3:

- Quad Core 1.2 GHz Broadcom BCM2837 64-bit CPU
- 1 GB RAM
- BCM43438 wireless LAN and **Bluetooth Low Energy (BLE)** on board
- 40 extended GPIO pins
- 4 USB 2 ports
- 4 Pole stereo output and composite video port
- Full-size HDMI
- CSI camera port for connecting a Raspberry Pi camera
- DSI display port for connecting a Raspberry Pi touchscreen display
- Micro SD port for loading your operating system and storing data
- Upgraded switched Micro USB power source up to 2.5 A

You will learn more technical details about Raspberry Pi in `Chapter 2`, *Building Your First Blynk Application*.

Raspberry Pi can be plugged in to a computer monitor or TV, and uses a standard keyboard and mouse. If you want, you can use the Raspberry Pi without attaching a monitor/TV, keyboard, or mouse.

You will need the following things to prepare your Raspberry Pi before installing Raspbian:

- **SD card**: 8 GB class 4 SD card is recommended; however, you can use one with higher capacity like 16 GB, 32 GB, or 64 GB. But having a class 10 micro SD card in your Raspberry Pi is very much recommended. Don't use SD cards lower than class 4, because they provide very low writing speed. The following screenshot shows an 8 GB Kingston class 4 micro SD card. The SD card works as the hard drive of your Raspberry Pi:

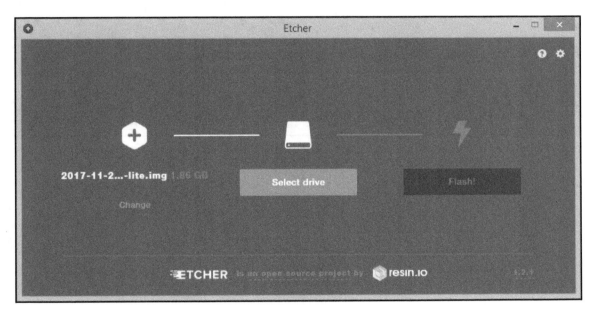

Kingston 8 GB class 4 micro SD card

- **Power supply**: Raspberry Pi is powered by a micro USB power supply. You'll need a good-quality power supply that can supply at least 2 A at 5V for the Raspberry Pi 3 Model B, or 700 mA at 5V for the earlier, lower-powered models. The following screenshot shows a micro USB power supply for Raspberry Pi:

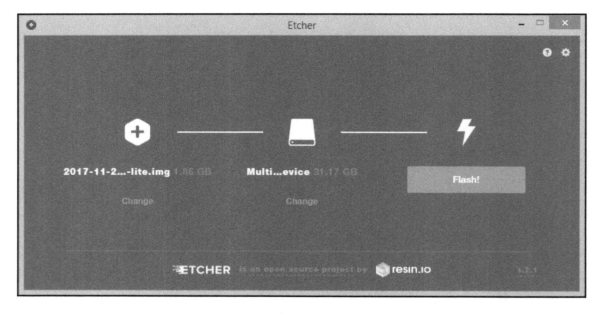

A micro USB power supply for Raspberry Pi

- **A Wi-Fi dongle (required for earlier, lower-powered models)**: Allows you to connect your Raspberry Pi to a wireless network. The following screenshot shows the official and recommended universal USB Wi-Fi dongle for Raspberry Pi.
- **Ethernet (network) cable [Model B/B+/2/3 only]**: An Ethernet cable is used to connect your Pi to a local network and the internet. The following screenshot shows an Ethernet cable:

Device List

Index	Computer Name	MAC Address	IP Address	Lease Time	Status	Type	Operation
1	Pradeeka	9C:2A:70:C 3:69:3F	192.168.1.2	0 days 12 hours 12 minutes 15 seconds	Active	Wi-Fi	Kick Out
2	raspberrypi	B8:27:EB:2 F:02:10	192.168.1.5	0 days 23 hours 57 minutes 5 seconds	Active	Ethernet	Kick Out
3	Unknown	90:23:EC:9 9:C7:57	192.168.1.3	0 days 22 hours 25 minutes 46 seconds	Inactive	Ethernet	Kick Out
4	android-f545dbea05 72d0c9	94:FE:22:6 5:B3:47	192.168.1.4	0 days 20 hours 45 minutes 26 seconds	Inactive	Ethernet	Kick Out

SparkFunCAT 6 Cable—3ft (image is CC BY 2.0—https://creativecommons.org/licenses/by/2.0/)

Setting up Raspberry Pi

Let's see how to set up Raspberry Pi:

1. Place your SD card into the SD card slot on the Raspberry Pi. It will only fit one way
2. Plug your keyboard and mouse in to the USB ports on the Raspberry Pi
3. Connect your HDMI cable from your Raspberry Pi to your monitor or TV
4. Connect a Wi-Fi dongle to one of the USB ports (unless you have a Raspberry Pi 3)
5. The Raspberry Pi is powered by the micro USB power supply

Setting up software on Raspberry Pi

The SD card works as the hard drive of your Raspberry Pi. As you know, every computer should have an operating system to manage.

The official operating system for Raspberry Pi is known as Raspbian. The Raspbian OS can be installed on the SD card.

Installing Raspbian

Raspbian is the officially supported operating system for Raspberry Pi by the Raspberry Pi Foundation. Raspbian can be installed on Raspberry Pi through NOOBS, or as a standalone image:

1. Browse the page `https://www.raspberrypi.org/downloads/` on the Raspberry Pi website. You can find two options at the top of the page:
   - **New Out Of Box Software** (**NOOBS**): An easy operating system installer for beginners
   - **Raspbian**: The officially supported Raspberry Pi operating system based on Debian

2. Click **Raspbian**:

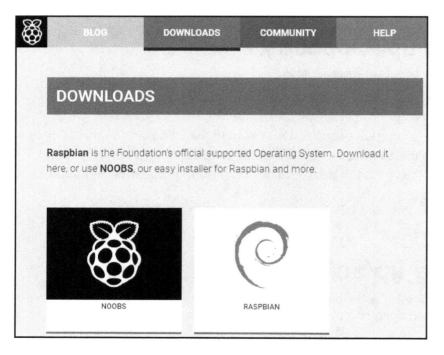

Raspbian

3. You will navigate to the `https://www.raspberrypi.org/downloads/raspbian/` page. Click on **RASPBIAN STRETCH LITE**, which is a **minimal image based on Debian Stretch**:

RASPBIAN STRETCH LITE

4. You can directly download the image file as a Zip file, or you can download it through the torrent. I found that downloading through the torrent is the fastest and most convenient way, but you will need a torrent client installed on your computer. uTorrent and BitTorrent are a couple of the popular clients. You can find plenty of resources on the internet on how to install and use torrent software.

5. Extract the downloaded compressed image file using 7-Zip (Windows), or the Unarchiver (macOS). You will get an image file like `2017-11-29-raspbian-stretch-lite.img`. The first part of the filename indicates the date on which you downloaded the file.

Writing Raspbian Stretch Lite image on SD card

You will need to use an image writing tool to install the image you have downloaded on your SD card.

Etcher is a graphical SD card writing tool that works on macOS, Linux, and Windows, and is the easiest option for most users. Etcher also supports writing images directly from the ZIP file, without any unzipping required. You can download Etcher at `https://etcher.io/`:

1. From the drop down-list, choose the operating system you are going to use to run Etcher (I chose Etcher for Windows x64):

Choosing the operating system

2. The executable file will start downloading, and will be saved to your computer's `Downloads` folder. You can use Etcher on your computer without installing it. Just run the executable file (`Etcher-Portable-1.2.1-x64.exe`) to launch the application:

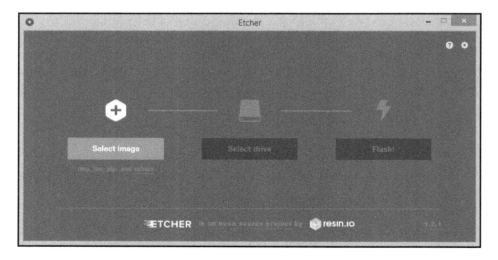

Etcher application window

3. Connect the micro SD card to your computer. You can use an SD card reader to connect it to your computer. Once connected, Windows will recognize it as a removable drive.

4. Click the **Select Image** button, browse the Raspbian image file (`2017-11-29-raspbian-stretch-lite.img`), and click the **Open** button. The following screenshot shows the resultant window after selecting the image file:

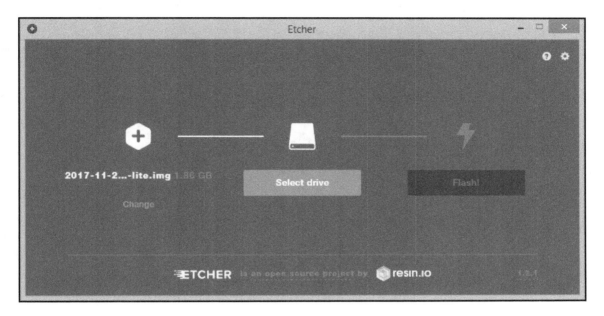

Selecting image

5. Click the **Select Drive** button and choose the SD card. The following screenshot shows the resultant window after selecting the drive for the SD card:

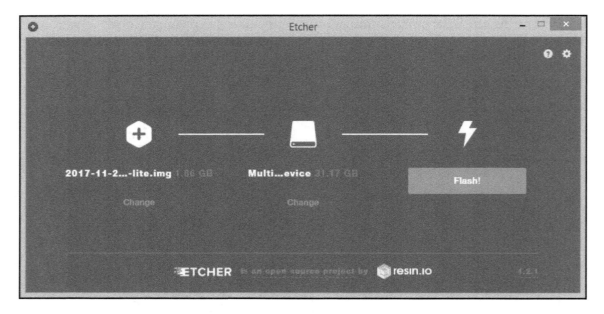

Selecting SD card as the drive

6. Click the **Flash!** button to write the image on your micro SD card. The following screenshot shows the window during flashing. When you get a message from Windows User Account Control, click the **Yes** button to proceed:

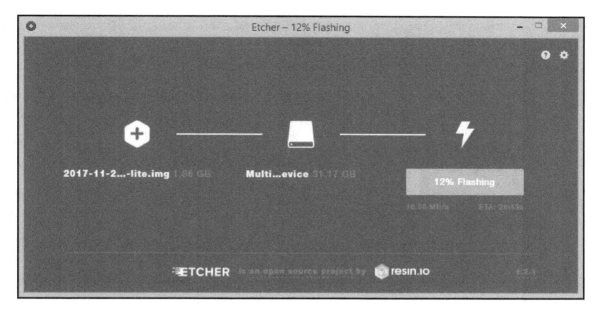

Image flashing to SD card in progress

Creating a configuration file for SSH

The Raspbian Stretch Lite doesn't have any configuration settings enabled to work with the SSH protocol for secure remote login from another computer. Therefore, first you should create a configuration file to enable the use of the SSH protocol:

1. With the SD card connected to your computer, open Command Prompt and change to your Pi's boot partition (let's assume the boot partition of the SD card is i):

   ```
   C:>i
   ```

2. Then type the following command to create an empty file named .ssh on the root of the boot partition of the SD card:

   ```
   i:>echo .ssh
   ```

3. Remove the SD card from your computer, and place it into the SD card slot on the Raspberry Pi. It will only fit one way.

4. Connect the USB Wi-Fi dongle to one of the USB ports (unless you have a Raspberry Pi 3).

5. Connect the Ethernet cable between your Raspberry Pi and the router.

6. Finally, connect the power supply to the micro USB port of the Raspberry Pi, and apply power. You will see a steady red LED and intermittently flashing green LED. Wait until the green LED turns off.

7. Log into your router's portal and find the dynamic IP address assigned with your Raspberry Pi:

Device List

Index	Computer Name	MAC Address	IP Address	Lease Time	Status	Type	Operation
1	Pradeeka	9C:2A:70:C 3:69:3F	192.168.1.2	0 days 12 hours 12 minutes 15 seconds	Active	Wi-Fi	Kick Out
2	raspberrypi	B8:27:EB:2 F:02:10	192.168.1.5	0 days 23 hours 57 minutes 5 seconds	Active	Ethernet	Kick Out
3	Unknown	90:23:EC:9 9:C7:57	192.168.1.3	0 days 22 hours 25 minutes 46 seconds	Inactive	Ethernet	Kick Out
4	android-f545dbea05 72d0c9	94:FE:22:6 5:B3:47	192.168.1.4	0 days 20 hours 45 minutes 26 seconds	Inactive	Ethernet	Kick Out

Raspberry Pi can be found under Device List

8. You will notice that the allocated IP address for my Raspberry Pi is `192.168.1.5`. You may get a different IP address based on your network topology.

Connecting with Raspberry Pi with SSH

SSH allows you to secularly log into the Raspberry Pi from your computer. PuTTY can be used to establish the connection between your computer and Raspberry Pi, and provides a terminal to run commands on Raspberry Pi. You can download PuTTY at `http://www.chiark.greenend.org.uk/~sgtatham/putty/download.html`.

The following steps will show you how to use PuTTY on Windows:

1. Open PuTTY by running the downloaded executable file (`putty.exe`).
2. In the **PuTTY Configuration** window, type in your connection settings:
   - **Host <u>N</u>ame (or IP address)**: IP address assigned by your router like `192.18.1.5`
   - **<u>P</u>ort**: `22`
   - **Connection type**: `SSH`
3. Then click the **Open** button to start the SSH session:

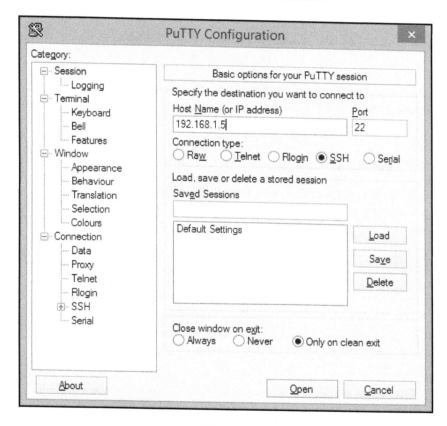

PuTTY configuration

4. If this is your first time connecting to Raspberry Pi from your computer, you will see the PuTTY security alert. Accept the connection by clicking the **Yes** button:

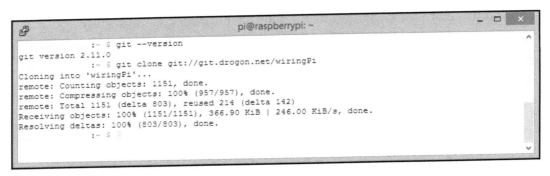

PuTTY Security Alert

5. Once the SSH connection is open, you should see a terminal. Type in the following default credentials to log into your Raspberry Pi. Press *Enter* following each input:

```
login as: pi
pi@192.168.1.5's password: raspberry
```

6. You are now logged in to your Raspberry Pi with SSH.

7. When typing your password, you will not see the cursor moving, or any characters typed on your terminal screen:

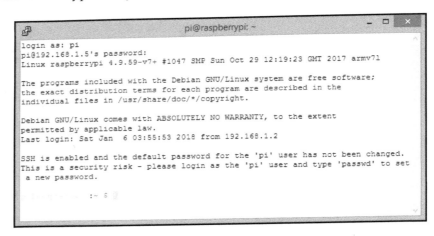

8. Log into Raspberry Pi using your default username and password.

Configuring a wireless connection on Raspberry Pi

Now you can configure the wireless connection to allow your Raspberry Pi to connect with your home or school Wi-Fi network. The following steps will guide you through how to configure the connection settings:

1. Open the `wpa_supplicant` configuration file in nano using the following command:

   pi@raspberrypi:~ $sudonano /etc/wpa_supplicant/wpa_supplicant.conf

2. Go to the bottom of the file and add the following:

   ```
   network={
   ssid="testing"
   psk="testingPassword"
   }
   ```

3. If the network you are connecting to does not use a password, the `wpa_supplicant` entry for the network will need to include the correct `key_mgmt` entry:

   ```
   network={
   ssid="testing"
   key_mgmt=NONE
   }
   ```

4. Alternatively, you can use the `wpa_passphrase` utility to generate an encrypted PSK:

   ```
   network={
   ssid="testing"
   #psk="testingPassword"
   psk=131e1e221f6e06e3911a2d11ff2fac9182665c004de85300f9cac208a6a
   80531
   }
   ```

5. If you are using a hidden network, an extra option in the `wpa_supplicant` file, `scan_ssid`, may help connection:

```
network={
ssid="yourHiddenSSID"
scan_ssid=1
psk="Your_wifi_password"
}
```

6. Now save the file by pressing *Ctrl + X*, then *Y*, then finally press *Enter*.

7. Reboot your Raspberry Pi for the new configurations to take effect by issuing the following command:

pi@raspberrypi:~ $sudo reboot

8. You can verify if it has successfully connected by using `ifconfig wlan0`. If the `inetaddr` field has an address beside it, the Raspberry Pi has connected to the network. If not, check that your `ssid` and `psk` are correct:

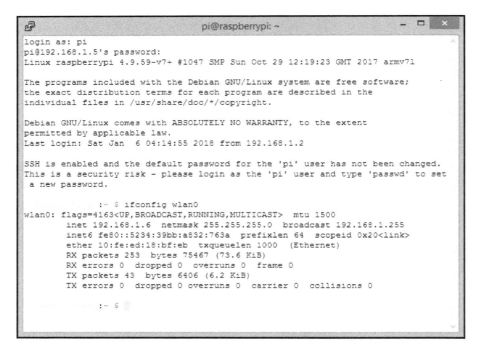

Verifying the wireless connection

9. Otherwise, you can verify the wireless connection between your Raspberry Pi and router by logging into your router's configuration page.

 You may get a different user interface depending on your router's brand.

Refer to the following screenshot:

Device List

Index	Computer Name	MAC Address	IP Address	Lease Time	Status	Type	Operation
1	Pradeeka	9C:2A:70:C3:69:3F	192.168.1.2	0 days 21 hours 59 minutes 25 seconds	Active	Wi-Fi	Kick Out
2	raspberrypi	B8:27:EB:2F:02:10	192.168.1.5	0 days 22 hours 49 minutes 47 seconds	Active	Ethernet	Kick Out
3	Unknown	90:23:EC:9:C7:57	192.168.1.3	0 days 20 hours 12 minutes 55 seconds	Inactive	Ethernet	Kick Out
4	android-f545dbea05 72d0c9	94:FE:22:65:B3:47	192.168.1.4	0 days 18 hours 32 minutes 35 seconds	Inactive	Ethernet	Kick Out
5	raspberrypi	10:FE:ED:18:BF:EB	192.168.1.6	0 days 22 hours 49 minutes 47 seconds	Inactive	Wi-Fi	Kick Out

Device list at router configuration page

10. Now you have successfully connected your Raspberry Pi to the home router's wireless network.

Installing prerequisite software on Raspbian

Now you're ready to install all the required software on your Raspberry Pi to prepare it to work with Blynk Cloud:

- Update and upgrade packages
- git core
- WiringPi
- Blynk libraries

Updating and upgrading Raspbian

Let's see how to update and upgrade Raspbian:

1. First, update your system's package list by entering the following command:

   ```
   pi@raspberrypi:~ $sudo apt-get update
   ```

   Refer to the following screenshot:

```
                    :~ $ sudo apt-get update
Get:1 http://mirrordirector.raspbian.org/raspbian stretch InRelease [15.0 kB]
Get:2 http://archive.raspberrypi.org/debian stretch InRelease [25.3 kB]
Get:3 http://mirrordirector.raspbian.org/raspbian stretch/main armhf Packages [1
1.7 MB]
Get:4 http://archive.raspberrypi.org/debian stretch/main armhf Packages [127 kB]
Get:5 http://mirrordirector.raspbian.org/raspbian stretch/contrib armhf Packages
 [56.8 kB]
Get:6 http://mirrordirector.raspbian.org/raspbian stretch/non-free armhf Package
s [95.2 kB]
Fetched 12.0 MB in 37s (319 kB/s)
Reading package lists... Done
                    :~ $
```

Updating Raspbian

2. Then, upgrade all your installed packages to their latest versions with the following command:

   pi@raspberrypi:~ $sudo apt-get dist-upgrade

3. When prompted, press *Y* to continue:

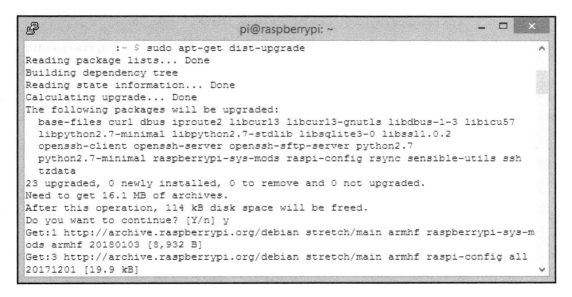

Upgrading Raspbian

Installing git

You will need to install git-core on your Raspberry Pi to work with git repositories for cloning packages like WiringPi, and Blynk libraries.

1. With the SSH to your Raspberry Pi with PuTTY, run the following command:

pi@raspberrypi:~ $sudo apt-get install git-core

```
                    :~ $ sudo apt-get install git-core
Reading package lists... Done
Building dependency tree
Reading state information... Done
The following additional packages will be installed:
  git git-man liberror-perl
Suggested packages:
  git-daemon-run | git-daemon-sysvinit git-doc git-el git-email git-gui gitk gitweb git-arch
  git-cvs git-mediawiki git-svn
The following NEW packages will be installed:
  git git-core git-man liberror-perl
0 upgraded, 4 newly installed, 0 to remove and 0 not upgraded.
Need to get 4,841 kB of archives.
After this operation, 26.3 MB of additional disk space will be used.
Do you want to continue? [Y/n] y
Get:4 http://mirrordirector.raspbian.org/raspbian stretch/main armhf git-core all 1:2.11.0-3+deb9u
2 [1,410 B]
Get:1 http://mirror.ossplanet.net/raspbian/raspbian stretch/main armhf liberror-perl all 0.17024-1
 [26.9 kB]
```

Installing git-core

2. This will install git on your Raspberry Pi. After installation is complete, verify the version of git by using the following command:

pi@raspberrypi:~ $ git -version

Refer to the following screenshot:

```
                    :~ $ git --version
git version 2.11.0
                    :~ $
```

Verifying git version

Installing WiringPi

WiringPi is a PIN-based GPIO access library that can be used with all versions of Raspberry Pi:

1. First, run the following command to clone the WiringPi repository:

   pi@raspberrypi:~ $ git clone git://git.drogon.net/wiringPi

   Refer to the following screenshot:

```
                  :~ $ git --version
git version 2.11.0
                  :~ $ git clone git://git.drogon.net/wiringPi
Cloning into 'wiringPi'...
remote: Counting objects: 1151, done.
remote: Compressing objects: 100% (957/957), done.
remote: Total 1151 (delta 803), reused 214 (delta 142)
Receiving objects: 100% (1151/1151), 366.90 KiB | 246.00 KiB/s, done.
Resolving deltas: 100% (803/803), done.
                  :~ $
```

Cloning WiringPi

2. Then move to the wiringPi directory:

   pi@raspberrypi:~ $ cd wiringPi

3. Run the following command to build the WiringPi:

   pi@raspberrypi:~/wiringPi $ ./build

Refer to the following screenshot:

```
                   :~ $ ls
wiringPi
                   :~ $ clear
                   :~ $ cd wiringPi
                   :~/wiringPi $ ./build
wiringPi Build script
========================


WiringPi Library
[UnInstall]
[Compile] wiringPi.c
[Compile] wiringSerial.c
[Compile] wiringShift.c
[Compile] piHiPri.c
[Compile] piThread.c
[Compile] wiringPiSPI.c
[Compile] wiringPiI2C.c
[Compile] softPwm.c
[Compile] softTone.c
[Compile] mcp23008.c
[Compile] mcp23016.c
[Compile] mcp23017.c
                 warning: `                     ' defined but not used [-Wunused-function]
 static        void digitalWrite8Dummy       (UNU struct wiringPiNodeStruct *node, UNU int pin, U
NU int value) { return ; }
                 warning: `                     ' defined but not used [-Wunused-function]
 static unsigned int digitalRead8Dummy       (UNU struct wiringPiNodeStruct *node, UNU int UNU pi
n)              { return 0 ; }

[Compile] mcp23s08.c
```

Building WiringPi

Deploying Blynk libraries

Let's look at how to deploy Blynk libraries:

1. You can deploy the Blynk libraries on Raspberry Pi by running the following commands:

```
pi@raspberrypi:~/wiringPi $ cd
pi@raspberrypi:~ $ git clone
https://github.com/blynkkk/blynk-library.git
```

Refer to the following screenshot:

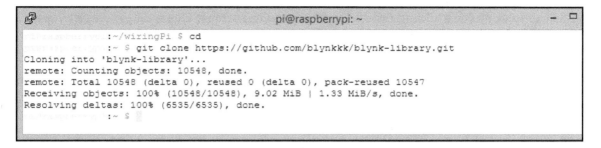

:~/wiringPi $ cd
:~ $ git clone https://github.com/blynkkk/blynk-library.git
Cloning into 'blynk-library'...
remote: Counting objects: 10548, done.
remote: Total 10548 (delta 0), reused 0 (delta 0), pack-reused 10547
Receiving objects: 100% (10548/10548), 9.02 MiB | 1.33 MiB/s, done.
Resolving deltas: 100% (6535/6535), done.
:~ $

Cloning Blynk libraries

2. After cloning the git repository, move to the `linux` directory:

   pi@raspberrypi:~ $ cd blynk-library/linux

3. Then build the library from source by targeting Raspberry Pi:

   pi@raspberrypi:~/blynk-library/linux $make clean all target=raspberry

4. This command will output the Blynk file that will be used to connect to the Blynk server:

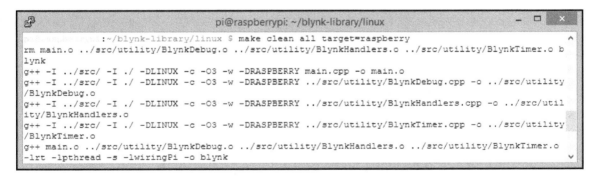

pi@raspberrypi: ~/blynk-library/linux

:~/blynk-library/linux $ make clean all target=raspberry
rm main.o ../src/utility/BlynkDebug.o ../src/utility/BlynkHandlers.o ../src/utility/BlynkTimer.o blynk
g++ -I ../src/ -I ./ -DLINUX -c -O3 -w -DRASPBERRY main.cpp -o main.o
g++ -I ../src/ -I ./ -DLINUX -c -O3 -w -DRASPBERRY ../src/utility/BlynkDebug.cpp -o ../src/utility/BlynkDebug.o
g++ -I ../src/ -I ./ -DLINUX -c -O3 -w -DRASPBERRY ../src/utility/BlynkHandlers.cpp -o ../src/utility/BlynkHandlers.o
g++ -I ../src/ -I ./ -DLINUX -c -O3 -w -DRASPBERRY ../src/utility/BlynkTimer.cpp -o ../src/utility/BlynkTimer.o
g++ main.o ../src/utility/BlynkDebug.o ../src/utility/BlynkHandlers.o ../src/utility/BlynkTimer.o -lrt -lpthread -s -lwiringPi -o blynk

Building Blynk libraries

Connecting Raspberry Pi with Blynk Cloud

After completing the building process, you can run the sample C++ code to connect your Raspberry Pi with the Blynk Cloud through Wi-Fi:

1. Run the default source code file, which is `main.cpp`, with the following command:

   pi@raspberrypi:~/blynk-library/linux $sudo ./blynk --token=da999e8ef4ac42148f9f e8427d482451

2. When you run the command `sudo ./blynk --token=da999e8ef4ac42148f9fe8427d482451`, the `main.cpp` file residing in the `blynk-library/linux` directory gets executed. The sample code only connects to the Blynk server at `https://www.blynk.cc/`. The output should look something like this:

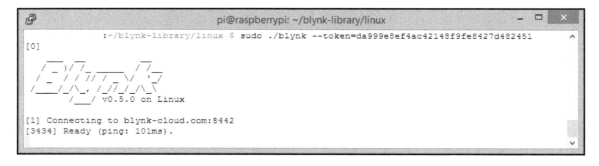

Output for default file main.cpp

3. You can find the sample code for `main.cpp` in the `blynk-library/linux` directory. You can read the content of the file with the following command:

   pi@raspberrypi:~/blynk-library/linux $ cat main.cpp

4. The output should look something similar to what is on this link: `https://github.com/PacktPublishing/Hands-On-Internet-of-Things-with-Blynk/blob/master/Chapter%201/Listing%201-1/main.cpp`.

5. Go through the code and see whether you can understand the key functions. Don't worry, you will learn how to write C++ code for the Blynk application in Chapter 3, *Using Controller Widgets*.

6. Now exit from the nano editor by pressing *Ctrl + X* with your keyboard. The editor will prompt you asking if you want to save the file; type Y in response.

Summary

Now you have to prepared your Raspberry Pi by equipping it with hardware and installing Raspbian OS and required software like git-core, WiringPi, and Blynk libraries. Finally, you ran the default C++ source file to connect your Raspberry Pi to the Blynk Cloud server.

Chapter 2, *Building Your First Blynk Application*, will present how to build your first application with Blynk. You will learn how to use the Blynk app builder to build an app to turn an LED on and off, and write C++ source code to allow your Raspberry Pi to connect, transfer, and manipulate data with Blynk Cloud.

2
Building Your First Blynk Application

Now you know how to connect your Raspberry Pi to the Blynk cloud with a simple C++ application running on itself. You also know how to create a project with Blynk app builder running on your smartphone or tablet to connect with your Raspberry Pi through the Blynk cloud.

In this chapter, we will cover the following:

- Attaching an LED to the Raspberry Pi
- Building an application with Blynk app builder
- Using GPIO_BCM pins with the Button widget
- Using virtual pins with the Button widget
- Writing C++ code with nano editor
- Editing the `build.sh` file
- Controlling the LED on the Pi by running the Blynk application (the C++ project and the Blynk app)

Controlling an LED

You are now ready to build your first Blynk application that can be used to control an LED attached to the Raspberry Pi from your smartphone or tablet. The application consists of the following components:

- Blynk app running on your smartphone or tablet
- C++ project running on Raspberry Pi
- Blynk cloud

Things you need

You will need following things to build the circuit:

- Raspberry Pi 3 board
- A breadboard
- An LED - Basic Red 5 mm (https://www.sparkfun.com/products/9590)
- A 330 Ohm 1/6 Watt PTH resistor (https://www.sparkfun.com/products/11507) color coded orange, brown, gold
- Two male-female jumper wires

Building the circuit

The following diagram shows how you can connect an LED to the BCM_GPIO pin 18 (physical pin 12) of the Raspberry Pi 3:

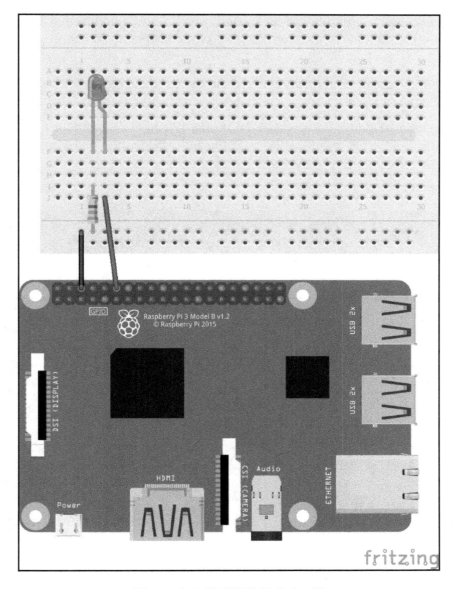

LED connected to the BCM_GPIO 18 of the Raspberry Pi 3

Connect the positive leg (the longer leg) of the LED to the physical pin 12 (BCM_GPIO PIN 18) of the Raspberry Pi.

Connect the negative leg (the shorter leg or the leg adjacent to the flat edge on the LED's plastic housing) of the LED to the physical pin 6 of the Raspberry Pi 3 through a 330 Ohm resistor.

Make sure that the positive and negative ends are properly seated on the breadboard. If not, the LED will not glow.

Following screenshot shows the physical, BCM, and WiringPi pin numbering table for the Raspberry Pi 3. Connect to your Raspberry Pi through SSH using PuTTY and run this command to get the pin numbering table:

pi@raspberrypi:~ $gpioreadall

```
pi@raspberrypi:~ $ gpio readall
+-----+-----+---------+------+---+---Pi 3---+---+------+---------+-----+-----+
| BCM | wPi |   Name  | Mode | V | Physical | V | Mode |   Name  | wPi | BCM |
+-----+-----+---------+------+---+----++----+---+------+---------+-----+-----+
|     |     |    3.3v |      |   |  1 || 2  |   |      | 5v      |     |     |
|   2 |   8 |   SDA.1 | ALT0 | 1 |  3 || 4  |   |      | 5V      |     |     |
|   3 |   9 |   SCL.1 | ALT0 | 1 |  5 || 6  |   |      | 0v      |     |     |
|   4 |   7 |  GPIO. 7|  IN  | 0 |  7 || 8  | 1 |  IN  | TxD     | 15  | 14  |
|     |     |      0v |      |   |  9 || 10 | 1 |  IN  | RxD     | 16  | 15  |
|  17 |   0 |  GPIO. 0|  IN  | 0 | 11 || 12 | 0 |  IN  | GPIO. 1 | 1   | 18  |
|  27 |   2 |  GPIO. 2|  IN  | 0 | 13 || 14 |   |      | 0v      |     |     |
|  22 |   3 |  GPIO. 3|  IN  | 0 | 15 || 16 | 0 |  IN  | GPIO. 4 | 4   | 23  |
|     |     |    3.3v |      |   | 17 || 18 | 0 |  IN  | GPIO. 5 | 5   | 24  |
|  10 |  12 |    MOSI | ALT0 | 1 | 19 || 20 |   |      | 0v      |     |     |
|   9 |  13 |    MISO | ALT0 | 1 | 21 || 22 | 0 |  IN  | GPIO. 6 | 6   | 25  |
|  11 |  14 |    SCLK | ALT0 | 0 | 23 || 24 | 1 |  OUT | CE0     | 10  | 8   |
|     |     |      0v |      |   | 25 || 26 | 1 |  OUT | CE1     | 11  | 7   |
|   0 |  30 |   SDA.0 |  IN  | 1 | 27 || 28 | 1 |  OUT | SCL.0   | 31  | 1   |
|   5 |  21 | GPIO.21 |  IN  | 0 | 29 || 30 |   |      | 0v      |     |     |
|   6 |  22 | GPIO.22 |  IN  | 0 | 31 || 32 | 0 |  IN  | GPIO.26 | 26  | 12  |
|  13 |  23 | GPIO.23 |  IN  | 1 | 33 || 34 |   |      | 0v      |     |     |
|  19 |  24 | GPIO.24 |  IN  | 0 | 35 || 36 | 0 |  IN  | GPIO.27 | 27  | 16  |
|  26 |  25 | GPIO.25 |  IN  | 0 | 37 || 38 | 0 |  IN  | GPIO.28 | 28  | 20  |
|     |     |      0v |      |   | 39 || 40 | 0 |  IN  | GPIO.29 | 29  | 21  |
+-----+-----+---------+------+---+----++----+---+------+---------+-----+-----+
| BCM | wPi |   Name  | Mode | V | Physical | V | Mode |   Name  | wPi | BCM |
+-----+-----+---------+------+---+---Pi 3---+---+------+---------+-----+-----+
```

Pin numbering table for Raspberry Pi 3

If you have an older version of Raspberry Pi, you can still build the project without including any extra components. The following diagram shows how you can connect an LED to the BCM_GPIO pin 18 (12 in physical pin numbering) of the Raspberry Pi 2:

LED connected to the GPIO 18 of the Raspberry Pi 2

The following screenshot shows the pin numbering table for Raspberry Pi 2 captured with the `gpioreadall` command:

BCM	wPi	Name	Mode	V	Physical	V	Mode	Name	wPi	BCM
		3.3v			1 || 2			5v		
2	8	SDA.1	ALTO	1	3 || 4			5V		
3	9	SCL.1	ALTO	1	5 || 6			0v		
4	7	GPIO. 7	IN	1	7 || 8	1	ALTO	TxD	15	14
		0v			9 || 10	1	ALTO	RxD	16	15
17	0	GPIO. 0	IN	0	11 || 12	0	IN	GPIO. 1	1	18
27	2	GPIO. 2	OUT	0	13 || 14			0v		
22	3	GPIO. 3	OUT	0	15 || 16	1	OUT	GPIO. 4	4	23
		3.3v			17 || 18	0	IN	GPIO. 5	5	24
10	12	MOSI	IN	0	19 || 20			0v		
9	13	MISO	IN	0	21 || 22	0	IN	GPIO. 6	6	25
11	14	SCLK	IN	0	23 || 24	1	IN	CE0	10	8
		0v			25 || 26	1	IN	CE1	11	7
0	30	SDA.0	IN	1	27 || 28	1	IN	SCL.0	31	1
5	21	GPIO.21	OUT	0	29 || 30			0v		
6	22	GPIO.22	IN	1	31 || 32	1	IN	GPIO.26	26	12
13	23	GPIO.23	IN	0	33 || 34			0v		
19	24	GPIO.24	IN	0	35 || 36	0	IN	GPIO.27	27	16
26	25	GPIO.25	IN	0	37 || 38	0	IN	GPIO.28	28	20
		0v			39 || 40	1	IN	GPIO.29	29	21
BCM	wPi	Name	Mode	V	Physical	V	Mode	Name	wPi	BCM

Pin numbering table for Raspberry Pi 2

The following diagram shows how you can connect an LED to the BCM_GPIO pin 18 (12 in physical pin numbering) of the Raspberry Pi 1:

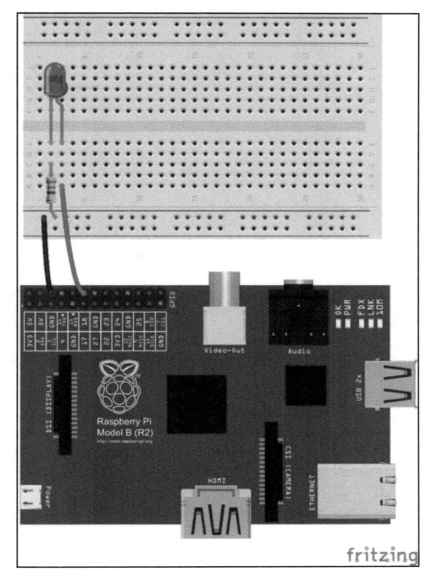

LED connected to the GPIO pin 18 of the Raspberry Pi 1

The following screenshot shows the pin numbering table for Raspberry Pi 1 captured with the `gpioreadall` command:

```
pi@raspberrypi ~ $ gpio readall
+-----+-----+---------+------+---+--B Plus--+---+------+---------+-----+-----+
| BCM | wPi |   Name  | Mode | V | Physical | V | Mode |  Name   | wPi | BCM |
+-----+-----+---------+------+---+----++----+---+------+---------+-----+-----+
|     |     |    3.3v |      |   |  1 || 2  |   |      | 5v      |     |     |
|   2 |   8 |   SDA.1 | ALTO | 1 |  3 || 4  |   |      | 5V      |     |     |
|   3 |   9 |   SCL.1 | ALTO | 1 |  5 || 6  |   |      | 0v      |     |     |
|   4 |   7 | GPIO. 7 |   IN | 0 |  7 || 8  | 1 | ALTO | TxD     | 15  | 14  |
|     |     |      0v |      |   |  9 || 10 | 1 | ALTO | RxD     | 16  | 15  |
|  17 |   0 | GPIO. 0 |   IN | 0 | 11 || 12 | 0 | IN   | GPIO. 1 | 1   | 18  |
|  27 |   2 | GPIO. 2 |   IN | 0 | 13 || 14 |   |      | 0v      |     |     |
|  22 |   3 | GPIO. 3 |   IN | 0 | 15 || 16 | 0 | IN   | GPIO. 4 | 4   | 23  |
|     |     |    3.3v |      |   | 17 || 18 | 0 | IN   | GPIO. 5 | 5   | 24  |
|  10 |  12 |    MOSI | ALTO | 0 | 19 || 20 |   |      | 0v      |     |     |
|   9 |  13 |    MISO | ALTO | 0 | 21 || 22 | 0 | IN   | GPIO. 6 | 6   | 25  |
|  11 |  14 |    SCLK | ALTO | 0 | 23 || 24 | 1 | OUT  | CE0     | 10  | 8   |
|     |     |      0v |      |   | 25 || 26 | 1 | OUT  | CE1     | 11  | 7   |
|   0 |  30 |   SDA.0 |   IN | 1 | 27 || 28 | 1 | IN   | SCL.0   | 31  | 1   |
|   5 |  21 | GPIO.21 |   IN | 1 | 29 || 30 |   |      | 0v      |     |     |
|   6 |  22 | GPIO.22 |   IN | 1 | 31 || 32 | 0 | IN   | GPIO.26 | 26  | 12  |
|  13 |  23 | GPIO.23 |   IN | 0 | 33 || 34 |   |      | 0v      |     |     |
|  19 |  24 | GPIO.24 |   IN | 0 | 35 || 36 | 0 | IN   | GPIO.27 | 27  | 16  |
|  26 |  25 | GPIO.25 |   IN | 0 | 37 || 38 | 0 | IN   | GPIO.28 | 28  | 20  |
|     |     |      0v |      |   | 39 || 40 | 0 | IN   | GPIO.29 | 29  | 21  |
+-----+-----+---------+------+---+----++----+---+------+---------+-----+-----+
| BCM | wPi |   Name  | Mode | V | Physical | V | Mode |  Name   | wPi | BCM |
+-----+-----+---------+------+---+--B Plus--+---+------+---------+-----+-----+
```

Pin numbering table for Raspberry Pi 1

Now, take your **Raspberry Pi universal power supply** and connect the micro USB plug to the micro USB connector of the Raspberry Pi.

You don't need a separate power supply to power the circuit. The Pi can provide enough power to drive the LED through the GPIO pin. The 330 Ohm resistor works as a current limiting resistor and protects the LED from high current flow.

Building the Blynk app

In this example, you will learn how to build a Blynk application to turn an LED on and off.

To start building a project, tap **Create New Project** in the center of the screen. This will bring you to the **Create New Project** wizard. You will walk through this wizard step by step.

Follow these steps to complete the Create New Project:

1. Tap **New Project**:

Creating a New Project

2. On the **Create New Project** page, type `LED Controller` in the **Project Name** textbox:

Providing a Project Name

3. Tap **CHOOSE DEVICE** to get the available hardware models and tap **Raspberry Pi 3 B** from the list. After choosing the hardware model, tap **OK**.

 If you have a previous version of Raspberry Pi board, choose the following:
Raspberry Pi 2: Raspberry Pi 2/A+/B+
Raspberry Pi 1 (Rev 1): Raspberry Pi B (Rev 1)
Raspberry Pi 1 (Rev 2): Raspberry Pi A/B (Rev 2)

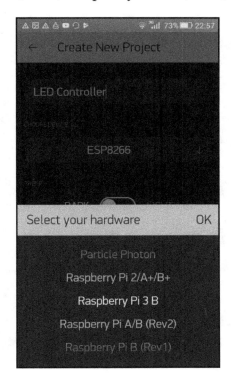

Selecting the hardware model

4. Tap **CONNECTION TYPE** to get the available connection types and tap **WiFi** from the list followed by the **OK** button.

5. Tap the **Create** button:

Creating a project

6. You will receive a new auth token in your email. You will need the auth token when you run the project.

Now, you have created the workspace for your project with all the required settings.

Adding a Button widget

These steps will show you how to add a Button widget to your project, which is a controller widget that can be used to turn an LED on and off.

The working area of the app builder is referred to as a canvas and has a dotted background with horizontal and vertical guides that allow you to easily arrange and align different types of widgets:

Canvas

1. Tap anywhere on the canvas or the plus icon on the toolbar to get the **Widget Box**. The **Widget Box** lists all the widgets available under different categories to build the application.
2. On the **Widget Box**, tap **Button** under **CONTROLLERS**:

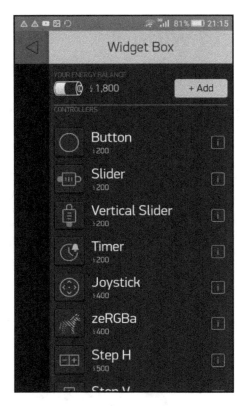

Button widget under CONTROLLERS

3. A Button widget will be added onto the canvas:

Button widget

4. Tap the Button widget to open the **Button Settings** page.
5. Tap **PIN** under **OUTPUT** and choose **Digital** followed by **GP18 PWM** from the list (this is the physical Pin 10/BCM Pin 18/GPIO Pin 18 *in* the GPIO header of the Raspberry Pi 3 connected to the LED).Then, tap the **OK** button.

6. Under **MODE**, slide the switch to the **SWITCH** position. This will configure the button as a switch:

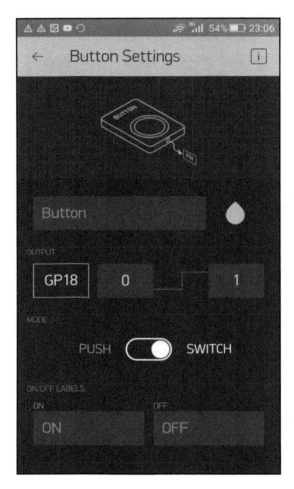

Configure the button as a switch

7. Tap the left arrow on the toolbar. Now, your configured Button widget should look as shown:

Button widget connected to gp18 (BCM 18)

8. Long tap and release the Button widget. You will see two points on the border of the Button widget:

Button widget with resizing points

9. Use these points to resize the Button widget by tapping and dragging on horizontal pane.

10. After resizing, tap on the canvas area to confirm the new dimension. Now your button should look as shown here:

Resized button

11. Long tap the Button widget until you see the move icon on the **Now** toolbar, and without releasing your finger, drag the button to a new location on the canvas:

Moving Button widget

12. The following screenshot shows the button moved to approximately the center of the canvas:

Button placed in about the center of the canvas

13. Now tap on the canvas to confirm the new location for the button:

Button moved to a new location

Now you have built the Blynk app using the Blynk app builder to control an LED connected to the Raspberry Pi.

Running the project

Now, you're ready to run your first Blynk project:

1. Log in to your Raspberry Pi using the SSH protocol with PuTTY (see `Chapter 1, Setting Up a Development Environment,` for information on logging in to your Raspberry Pi with PuTTY).

2. Change to the `blynk-library/linux` directory by issuing this command:

   pi@raspberrypi:~ $`cd blynk-library/linux`

3. Run the following command with the auth token associated with your project:

   pi@raspberrypi:~/blynk-library/linux $ `sudo ./blynk --token=ca7bed1c92214503a65de8e20164994f`

4. The terminal will show an output similar to this:

   ```
   [0]

   ___ __ __

   / _ )/ /_ ____ / /__
   / _ / / // / _ / '_/
   /____/_/_, /_//_/_/_
   /___/ v0.5.0 on Linux

   [1] Connecting to blynk-cloud.com:8442

   [3490] Ready (ping71ms).
   ```

Playing the app

Now, your Raspberry Pi is successfully connected with the Blynk cloud at `https://www.blynk.cc/`. The following steps show you how to use the app to control the LED attached to GPIO pin 18:

1. Tap the **Play** button on the toolbar:

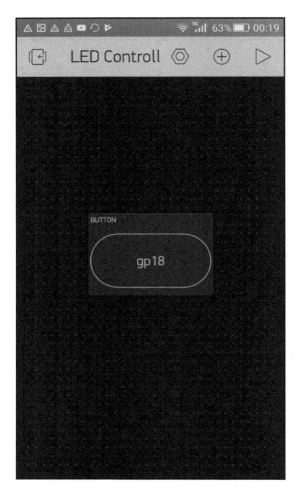

Tapping Play button on the toolbar

2. This will switch you from edit mode to play mode, where you can interact with the hardware. The play button is replaced with a **Stop** button that allows you to go back to edit mode:

Application in play mode

3. You can tap the chip icon (the icon like a microprocessor) to see the device status, whether it is online or offline. The following screenshot show that the Raspberry Pi is online:

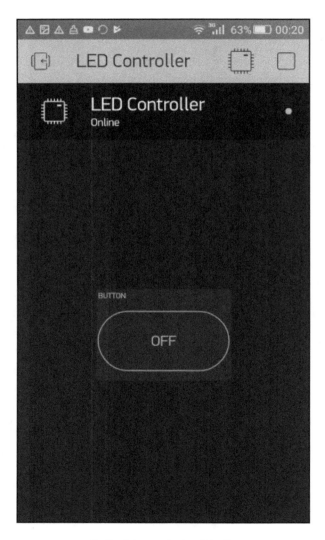

Checking device status: Raspberry Pi is online

4. If your Raspberry Pi is offline, the notification looks as shown in the following screenshot. There are several reasons it can go offline. Some of them are:

- Lost internet connection on router
- Lost Wi-Fi (or Ethernet) connection on Raspberry Pi
- Lost power on Raspberry Pi
- `main.cpp` is not executed
- There is a bug in `main.cpp`

Indicating the device is offline

5. Tap **BUTTON** to turn the LED on and off. The button will act as a switch and hold its state until you tap it again:

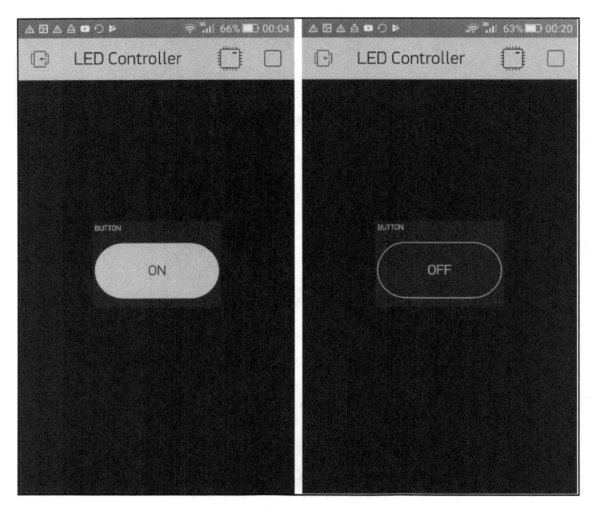

Button states for ON and OFF

6. You can tap the stop button and get back to edit mode.
7. While in play mode, you won't be able to drag or set up new widgets on the canvas.

Using virtual pins

Now you know how to write the current state of the Button widget to a GPIO pin.

With virtual pins, you can send or receive any type of data between the microcontroller (Raspberry Pi) and the Blynk app. The Blynk app builder provides 128 virtual pins for working with Raspberry Pi. They allow you to interface with sensors, actuators, and libraries. These pins have no physical properties. Some Blynk widgets support virtual pins, some do not.

As an example, you can write the current state of the Button widget to the virtual pin V1 as shown here. Then, you can assign the incoming value from the virtual pin to a variable for further processing:

```
BLYNK_WRITE(V1)

{

intpinValue = param.asInt(); // assigning value on Virtual Pin (V1) to a
variable

// process received value

}
```

Configuring Button widget with virtual pin

In this section, you will modify the Blynk app, LED Controller, by configuring a virtual pin:

1. Go to edit mode by tapping the stop icon if you're already in play mode.
2. Tap the Button widget.
3. On the **Button Settings** page, tap **GP 18** under **OUTPUT**. Scroll down and choose **Virtual** from the left-hand part of the list. Then, scroll down and choose **V1** from the right-hand part of the list under **PIN**. Blynk provides 128 virtual pins for working with Raspberry Pi. Finally, tap **OK**.

4. Now, your **Button Settings** page should look like this:

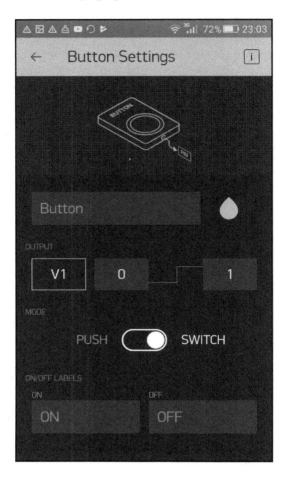

Button settings for virtual pin

5. Tap the **LEFT** arrow on the toolbar. You will see the new button configured with the virtual pin **V1**:

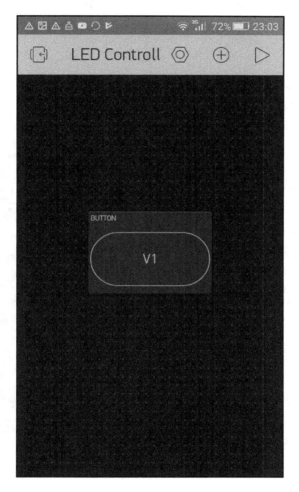

Button configured with virtual pin V1

Now, you've successfully modified the Button widget, so it can write values to the virtual pin **V1**.

Modifying the main.cpp file

To complete your application, the `main.cpp` file should also be modified to process the data on the virtual pin. These steps show you how to modify it with the nano editor:

1. Open the `main.cpp` file with nano by issuing the following command:

pi@raspberrypi:~/blynk-library/linux $ sudonanomain.cpp

main.cpp opened with nano editor

2. Before using the `WiringPi GPIO` library, you need to include its header file in your program as follows:

```
#include <wiringPi.h>
```

3. Scroll down to the file with the arrow keys on your keyboard and add the following code to the `setup()` function:

```
void setup()
{
Blynk.begin(auth, serv, port);
pinMode(1, OUTPUT);
}
```

The `pinMode()` function allows you to set the mode of a pin to either INPUT, OUTPUT, or PWM_OUTPUT. Note that only WiringPi pin 1 (BCM_GPIO 18) supports PWM output.

4. Then, add the following lines to the `BLYNK_WRITE(V1)` function. This code snippet will compare the value on virtual pin V1 and write values on BCM_GPIO 18 pin (WiringPi pin 1) as follows:

```
V1 BCM_GPIO 18 / WiringPi 1
1 HIGH
0 LOW
BLYNK_WRITE(V1)
{
printf("Got a value%sn", param[0].asStr());
if(param.asInt() == 1)
digitalWrite(1, HIGH);
else
digitalWrite(1, LOW);
}
```

5. Press *Ctrl + O* followed by *Enter* to save the changes.
6. Press *Ctrl + X* to *Exit* the nano editor.

Listing 2.1 shows the complete code for `main.cpp`:

Listing 2.1 Controlling an LED using Virtual Pins

```
/**

* @file main.cpp

* @author VolodymyrShymanskyy

* @license This project is released under the MIT License (MIT)
```

```
* @copyright Copyright (c) 2015 VolodymyrShymanskyy

* @date Mar 2015

* @brief

*/

//#define BLYNK_DEBUG

#define BLYNK_PRINT stdout

#ifdef RASPBERRY

#include <BlynkApiWiringPi.h>

#else

#include <BlynkApiLinux.h>

#endif

#include <BlynkSocket.h>

#include <BlynkOptionsParser.h>

static BlynkTransportSocket _blynkTransport;

BlynkSocket Blynk(_blynkTransport);

static const char *auth, *serv;

static uint16_t port;

#include <BlynkWidgets.h>

#include <wiringPi.h>

BLYNK_WRITE(V1)

{

printf("Got a value%sn", param[0].asStr());

if(param.asInt() == 1)

digitalWrite(1, HIGH);
```

```
else

digitalWrite(1, LOW);

}

void setup()

{

Blynk.begin(auth, serv, port);

pinMode(1, OUTPUT);

}

void loop()

{

Blynk.run();

}

int main(int argc, char* argv[])

{

parse_options(argc, argv, auth, serv, port);

setup();

while(true) {

loop();

}

return 0;

}
```

7. Open the `build.sh` file with nano by issuing this command:

pi@raspberrypi:~/blynk-library/linux $sudonano build.sh

8. Modify the file to exactly as shown here:

```
#!/bin/bash

case "$1" in

raspberry)

make clean all target=raspberry

exit 0

;;

linux)

make clean all

exit 0

;;

esac

echo "Please specify platform raspberry, linux"

exit 1
```

9. Press *Ctrl + O* followed by *Enter* to save the changes.
10. Press *Ctrl + X* to *Exit* the nano editor.
11. Build your C++ project by issuing this command:

pi@raspberrypi:~/blynk-library/linux $ ./build.sh raspberry

Running the project

Let's see how to run the project:

1. After building the project, run it with this command:

pi@raspberrypi:~/blynk-library/linux $sudo ./blynk --token=ca7bed1c92214503a65de8e20164994f

2. You should get an output that looks like this:

```
pi@raspberrypi: ~/blynk-library/linux

pi@raspberrypi:~/blynk-library/linux $ sudo ./blynk --token=ca7bed1c92214503a65d
e8e20164994f
[0]

   ___  __            __
  / _ )/ /_ _____  / /__
 / _  / / // / _ \/  '_/
/____/_/\_, /_//_/_/\_\
       /___/ v0.5.0 on Linux

[4] Connecting to blynk-cloud.com:8442
[3327] Ready (ping: 60ms).
```

Output on console when running the project

3. Tap the **Play** button on the toolbar. Now, you can turn on and off the LED by tapping the button. When you tap the button, the console will output the current value on the virtual pin V1.

4. The following screenshot shows the console output when you tap the Button widget:

```
pi@raspberrypi:~/blynk-library/linux $ sudo ./blynk --token=ca7bed1c92214503a65d
e8e20164994f
[1]

       ___  __          __
      / _ )/ /_ _____  / /__
     / _  / / // / _ \/  '_/
    /____/_/\_, /_//_/_/\_\
           /___/ v0.5.0 on Linux

[2] Connecting to blynk-cloud.com:8442
[3958] Ready (ping: 71ms).
Got a value: 1
Got a value: 0
Got a value: 1
Got a value: 0
Got a value: 1
Got a value: 0
Got a value: 1
Got a value: 0
Got a value: 1
Got a value: 0
Got a value: 1
```

Console output while tapping Button widget

Summary

You have learned a lot and built your very first complete Blynk application, which consists of a Blynk hardware and a Blynk app to control an LED attached to the Raspberry Pi over Wi-Fi from a smartphone or a tablet. Now you know how to add the Button widget and configure it to work with the physical pins of the hardware, as well as with virtual pins. You also know how to write code in C++ to process the values on the virtual pin for the Button widget.

In Chapter 3, *Using Controller Widgets*, you will learn in detail about all the available controller widgets that can be found in the Blynk app builder.

Using Controller Widgets

3

Controller widgets allow you to control any actuators attached to the Raspberry Pi through GPIO pins. The pins can be digital or virtual. With controller widgets you can, for example, control the brightness of an LED, change the color of an RGB LED, or change the speed of motors to control the movement of a robot buggy.

In this chapter, you will learn:

- About controller widgets such as Slider, Step, Joystick, and zeRGBa
- How to use software PWM library
- How to connect controller widgets with digital and virtual pins
- How to use split and merge mode
- How to parse values

Creating a project

First, create a new project as discussed in previous chapters, and name it `Controllers`. You will receive a new auth token. However, if you would like to use the workspace of the project LED controller you built in `Chapter 2`, *Building Your First Blynk Application*, just rename it `Controllers`. This will allow you to use the same auth token without applying a new one.

Use the following steps to clean and rename the project:

1. In edit mode, first delete the Button widget. To delete, tap the Button widget, then in the **Button Settings** page, tap **Delete**. When prompted, tap **OK** to continue. Now your canvas is empty:

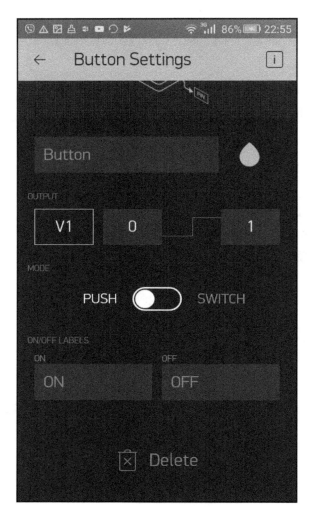

Deleting the Button widget

2. Then, tap the nut icon to open the **Project Settings**.
3. In the **Project Name** textbox, change the project name to `Controllers`. Scroll down the page and tap **LED Controller** under **Devices** to rename the device:

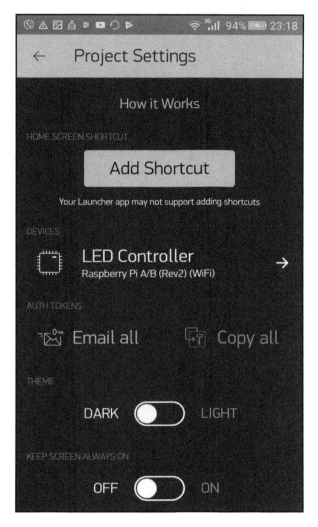

Renaming the project

4. Tap **LED Controller** under **My Devices**:

My devices

5. Change the device name to Controllers:

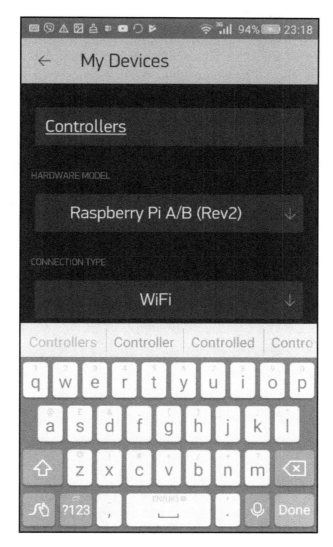

Renaming the device

6. Finally, tap the back icon three time to go to the canvas.

Slider

The Slider widget allows you to send a specific range of values from your Blynk app to the hardware (Raspberry Pi). It is analogous to a potentiometer in electronics. The default value of the Slider widget provides a value range of 0 to 255. As examples, sliders can be used to control the brightness of an LED or to control a servo motor.

In this example project, you will control the brightness of an LED attached to the Raspberry Pi BCM_GPIO pin 18, which is a **Pulse Width Modulation (PWM)** pin.

Brightness can be changed by analog or PWM. With analog, you can control the brightness by simply adjusting the DC current in the wire. But with PWM, you can change the brightness by varying the duty cycle of a constant current in the wire. Therefore, PWM can be implemented on any digital pin without using a potentiometer to control the current flow. PWM is the most appropriate method to change the brightness of the LED without any significant loss of accuracy, and no change in LED color.

This uses the same wiring diagram for Raspberry Pi 3, as shown in the diagram (see Chapter 2, *Building Your First Blynk Application*, for connecting an LED to previous versions of Raspberry Pi):

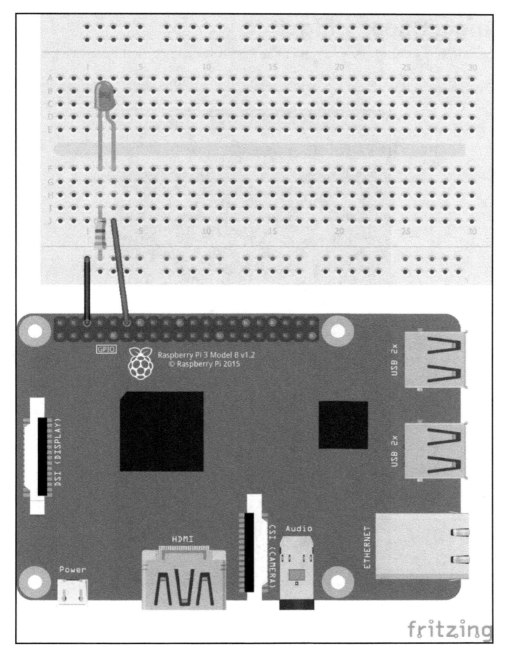

LED connected to the Raspberry Pi 3 BCM_GPIO 18

Adding a Slider widget

The following steps will show you how to add a **Slider** widget to the canvas:

1. Tap the plus icon on the toolbar to open the **Widget Box**:

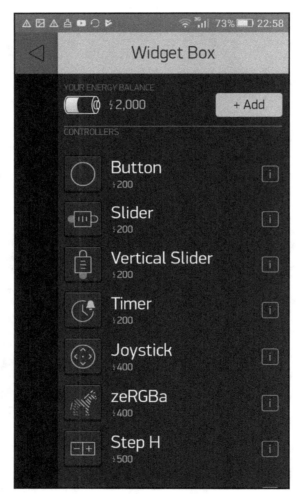

Choosing Slider widget from Widget Box

2. Tap **Slider**. A Slider widget will be added to the canvas. The Slider widget consumes 200 energy:

Slider widget placed on the canvas

3. This is a horizontal slider, but if you want to add a vertical slider, just tap vertical slider in the **Widget Box**. Except for the orientation, both have the same properties. The following figure shows a vertical slider added to the canvas:

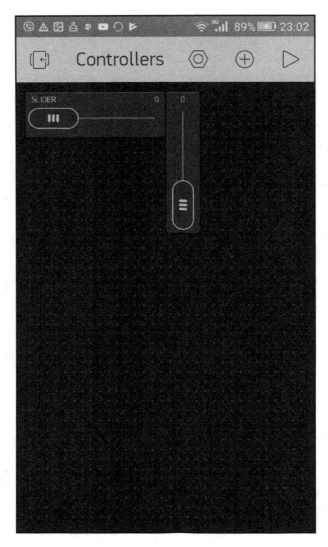

Vertical Slider (right)

Using digital pins

Let's configure the Slider widget to control the brightness of the LED using digital pins:

1. Tap the **Slider** widget to open the **Slider Settings** page:

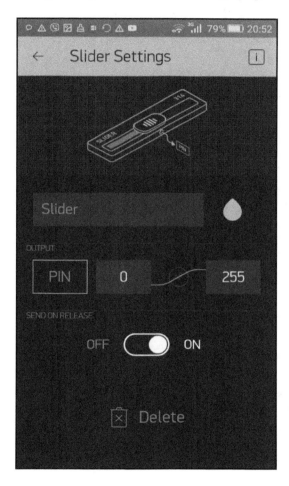

Slider Settings page

2. Tap **PIN** under **OUTPUT**. On the left-hand side of the popup, choose **Digital**, and on the right-hand side of the popup, under **PIN**, choose **gp 18 PWM** (this is the only pin available on Raspberry Pi for PWM). Then, tap **OK**:

Selecting a digital pin for PWM

3. Change the maximum value for the **OUTPUT** to 1023 by replacing the default value of 255. Now, your **Slider Settings** page should look as follows:

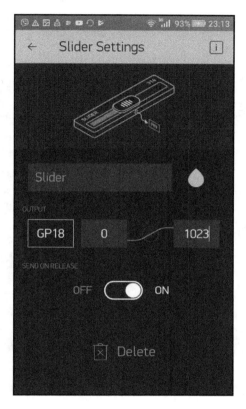

Modified Slider Settings page

4. Keep the **SEND ON RELEASE** toggle button in the **ON** position. This will only write a value on the Raspberry Pi GPIO 18 pin when you release the slider after moving to a new position. If you set **SEND ON RELEASE** to the **OFF** position, the Slider widget will write values on GPIO 18 while you're moving the slider. The write interval for that can be defined under **WRITE INTERVAL**. The minimum value for write interval is 100 ms.

5. Tap the back icon on the toolbar. Now, you should see the **GP18** label on the **SLIDER** with the minimum value it can take as 0:

Slider attached to BCM_GPIO 18 PWM

6. Connect to your Raspberry Pi with your computer using PuTTY. In the Terminal window, type the following commands to connect your Raspberry to the Blynk cloud:

```
pi@raspberrypi:~ $ cd blynk-library/linux
pi@raspberrypi:~/blynk-library/linux $sudo ./blynk --
token=ca7bed1c92214503a65de8e20164994f
```

7. Tap the play button to start the app. Initially, the slider is at 0 position and your LED should be off. Move the slider to a new position and release the fingertip. The LED will turn on and the brightness will be set to the new value of the slider. For your reference:

- 0: Off
- 1 to 1023: On
- 1023: Maximum brightness

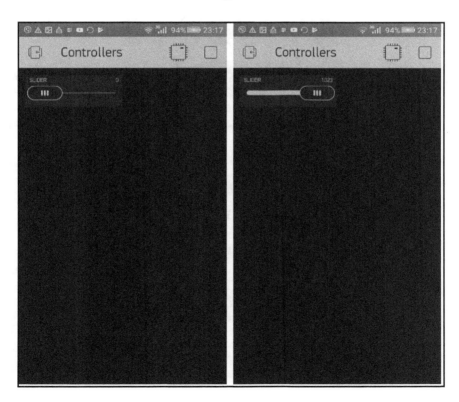

Changing Slider value from 0 (off) to 1023 (maximum brightness)

Using virtual pins

If you want to further process the values coming from the Slider widget, virtual pins are your best friends:

1. First, go to edit mode by tapping the stop icon on the toolbar.
2. Tap **GP18** under **OUTPUT**. Choose **Virtual** from the left-hand side list and **V1** from the right-hand side list (this is the only pin available on Raspberry Pi for PWM). Then, tap **OK**.
3. The maximum value for **OUTPUT** should be 1023.
4. Now, your **Slider Settings** page should look as follows:

Slider settings for working with Virtual Pi V1

5. Tap the back icon to go to the canvas.

6. Connect to your Raspberry Pi using PuTTY and open the `main.cpp` file using nano.

7. Modify the `setup()` function as follows. The mode of the BCM_GPIO pin 18 (WiringPi pin 1) is set as `PWM_OUTPUT` to configure it as a PWM pin:

```
void setup()
{
Blynk.begin(auth, serv, port);
pinMode(1, PWM_OUTPUT);
}
```

8. Modify the `BLYNK_WRITE(V1)` function as follow. The `pwmWrite()` function writes on BCM_GPIO pin 18 (WiringPi pin 1):

```
BLYNK_WRITE(V1) //Slider Widget is writing to pin V1
{
printf("Got a value: %sn", param[0].asStr());
intpinData = param.asInt();
pwmWrite(1, pinData);
}
```

9. Build the C++ project by issuing the following command:

```
pi@raspberrypi:~/blynk-library/linux $sudo ./build.sh
raspberry
```

10. Run the C++ project with this command:

```
pi@raspberrypi:~/blynk-library/linux $sudo ./blynk --
token=ca7bed1c92214503a65de8e20164994f
```

11. Then, start your Blynk app by tapping the play button on the toolbar. Drag the slider to a new position and release it. The brightness of the LED will change according to the slider value between 0 to 1023:

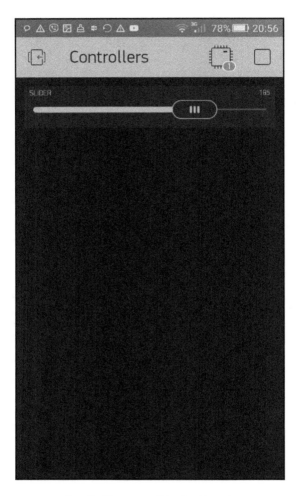

Controlling LED brightness Slider using a Virtual pin

Listing 3.1 shows the source code for `main.cpp`:

Listing 3.1: Controlling brightness of an LED with the Slider widget

```cpp
//#define BLYNK_DEBUG

#define BLYNK_PRINT stdout

#ifdef RASPBERRY

#include <BlynkApiWiringPi.h>

#else

#include <BlynkApiLinux.h>

#endif

#include <BlynkSocket.h>

#include <BlynkOptionsParser.h>

#include <wiringPi.h>

#include <softPwm.h>

static BlynkTransportSocket _blynkTransport;

BlynkSocket Blynk(_blynkTransport);

static const char *auth, *serv;

static uint16_t port;

#include <BlynkWidgets.h>

BLYNK_WRITE(V1)

{

printf("Got a value: %sn", param[0].asStr());

intpinData = param.asInt();

pwmWrite(1, pinData);
```

```
}

void setup()

{

Blynk.begin(auth, serv, port);

pinMode(1, PWM_OUTPUT);

}

void loop()

{

Blynk.run();

}

int main(intargc, char* argv[])

{

parse_options(argc, argv, auth, serv, port);

setup();

while(true) {

loop();

}

return 0;

}
```

Step

The Step widget consist of two buttons, one to increment the value and another one to decrement the value. You can predefine the range of values and the size of a step that the Step widget can take.

Adding a Step widget

These steps will show you how to add a Step widget to the canvas:

1. Tap the plus icon to open the **Widget Box**. Then, tap the **STEP H** widget.
2. A horizontal Step widget will be added to your canvas. The **STEP** widget consumes 500 energy:

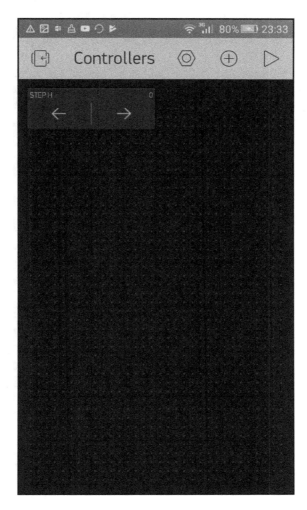

Horizontal Step widget

3. If you want to add a vertical Step widget, tap **STEP V** in the **Widget Box**:

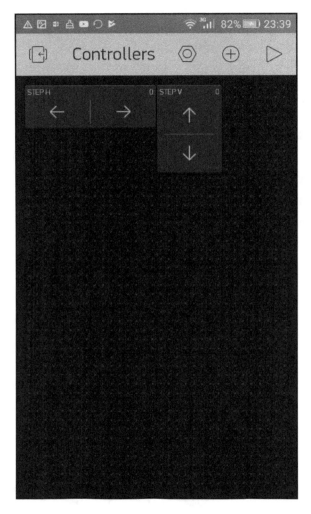

Step V (vertical step) widget

Using digital pins

Let's configure the Step widget to control the brightness of the LED attached to the Raspberry Pi BCM_GPIO pin 18 using digital pins:

1. Tap **Slider H** to open the **Step H Settings** page:

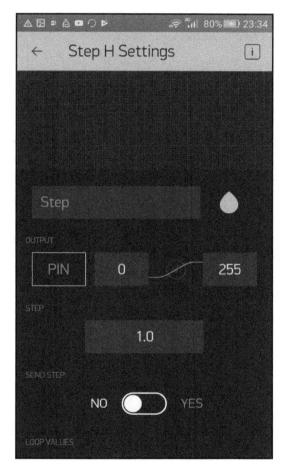

Step H Settings

2. Tap **PIN** under **OUTPUT**. Choose **Digital** from the left-hand list and **gp 18 PWM** from the right-hand list. Then, tap **OK**.

3. By default, **OUTPUT** ranges from 0 to 255. Under **OUTPUT**, tap **255** and change its value to 1023. This is the maximum range for PWM that can be used to control the brightness of an LED.

4. By default, the size of each step is 1.0. If you want to change the size of the step, tap the text box under **STEP** and type a new value.

5. Under **SEND STEP**, keep the toggle button to the **NO** position. If you move it to **YES**, the step value will be sent to the hardware pin instead of the actual value of the Step widget (in this case 1.0).

6. Under **LOOP VALUES**, keep the toggle button in the **OFF** position. The **LOOP VALUES** option allows you to reset the Step widget to the start value when the maximum value is reached.

7. You can customize the increment and decrement symbols on the Step widget by using the option under **ICONS**.

8. The writing interval for values can be selected under **WRITE INTERVAL**. The default value is 100 ms.

> **WRITE INTERVAL** allows you to send values to your Raspberry Pi within a certain interval. As an example, setting **WRITE INTERVAL** to 100 ms will send one value to the Raspberry Pi from a widget within a 100 ms period. This option allows you to optimize data traffic flow to your Raspberry Pi.

9. Tap the back arrow to go to the canvas.

10. Tap the play button to start the application.

11. Tap the slider to a new position. The LED attached to the BCM_GPIO pin 18 will change its brightness according to the value (from 0 to 1023) it receives.

Using virtual pins

Let's configure the Step widget to use virtual pins. These steps will show you how to do it:

1. In edit mode, tap the Step widget.

2. Tap **PIN** under **OUTPUT**. Choose **Virtual** from the left-hand list and **V1** from the right-hand list (this is the only pin available on the Raspberry Pi for PWM). Then, tap **OK**.

3. The maximum value under **OUTPUT** should be 1023.

4. After configuring all this, your **Step H Settings** page should look like this:

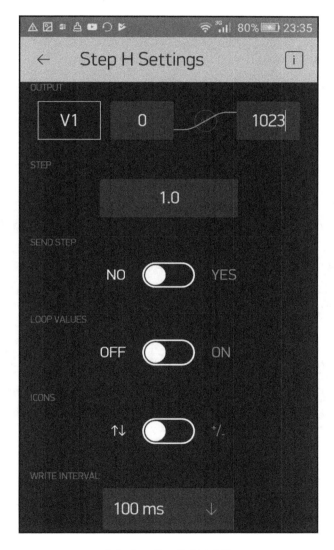

Settings for working with virtual pin V1

5. Tap the back icon on the toolbar to go to the canvas.

6. You can use the same C++ project you built in the previous section (Slider) without making any modifications. The Step widget uses virtual pin V1 to write values on WiringPi pin 1, which is a PWM pin.

7. Run the C++ project with this command:

   pi@raspberrypi:~/blynk-library/linux $sudo ./blynk --
   token=ca7bed1c92214503a65de8e20164994f

8. Then, start your Blynk app by tapping the play button on the toolbar. Tap the increase and decrease buttons to change the value between 0 and 1023. The brightness of the LED will change according to the value output by the **Step** widget:

Controlling brightness of an LED with Step widget

Listing 3.2 shows the source code for the main.cpp file:

Listing 3.2: Controlling brightness of an LED with Step H widget

```
//#define BLYNK_DEBUG

#define BLYNK_PRINT stdout

#ifdef RASPBERRY

#include <BlynkApiWiringPi.h>
```

```
#else

#include <BlynkApiLinux.h>

#endif

#include <BlynkSocket.h>

#include <BlynkOptionsParser.h>

#include <wiringPi.h>

#include <softPwm.h>

static BlynkTransportSocket _blynkTransport;

BlynkSocket Blynk(_blynkTransport);

static const char *auth, *serv;

static uint16_t port;

#include <BlynkWidgets.h>

BLYNK_WRITE(V1)

{

printf("Got a value: %sn", param[0].asStr());

intpinData = param.asInt();

pwmWrite(1, pinData);

}

void setup()

{

Blynk.begin(auth, serv, port);

pinMode(1, PWM_OUTPUT);

}

void loop()
```

```
{

Blynk.run();

}

intmain(intargc, char* argv[])

{

parse_options(argc, argv, auth, serv, port);

setup();

while(true) {

loop();

}

return 0;

}
```

zeRGBa

zeRGBa is a controller widget for picking colors by mixing red, green, and blue. Each basic color ranges from 0 to 255. The RGB color system constructs all the colors from the combination of the red, green and blue. The red, green, and blue use eight bits each, which have integer values from 0 to 255. This makes *256*256*256=16,777,216* possible colors.

The zeRGBa widget uses three pins on your hardware (Raspberry Pi) for each color channel: red, green, and blue.

Using digital pins

In split mode, the zeRGBa widget uses three pins on your hardware for each basic color. The pins should be capable of PWM. Unfortunately, the Raspberry Pi has a single pin that is GPIO 18 capable with PWM, but you will require two more PWM pins. Therefore, you can't use the zeRGBa widget with Raspberry Pi digital pins.

Using virtual pins

In merge mode, the zeRGBa widget writes a single message to the virtual pin, which consists of an array of values. As an example, if you have a zeRGBa widget connected to the virtual pin V1, you can parse each value for RGB on your Raspberry Pi as:

```
BLYNK_WRITE(V1) // zeRGBa assigned to V1

{

// get a RED channel value

int r = param[0].asInt();

// get a GREEN channel value

int g = param[1].asInt();

// get a BLUE channel value

int b = param[2].asInt();

}
```

The following steps explain how to add a zeRGBa widget to the canvas and configure it for use with virtual pins:

1. In edit mode, tap the plus icon to open the **Widget Box**.

2. Under **Controllers**, tap **zeRGBa**. A **zeRGBa** widget will be added to the canvas:

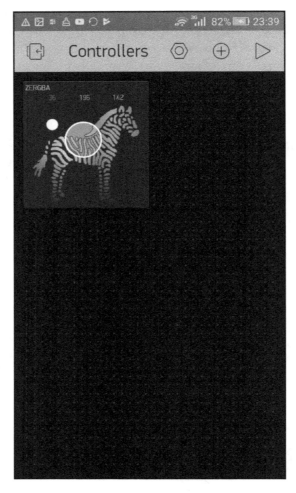

zeRGBa widget

3. Tap the **zeRGBa** widget to open **zeRGBa Settings**:

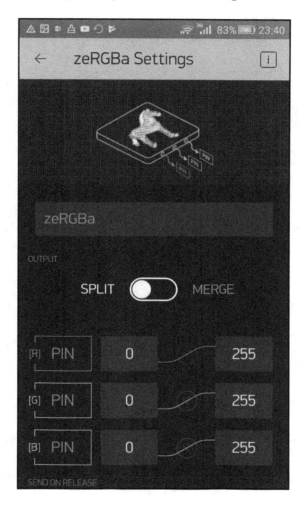

zeRGBa Settings page

4. Switch the toggle button under **OUTPUT** to the **MERGE** position:

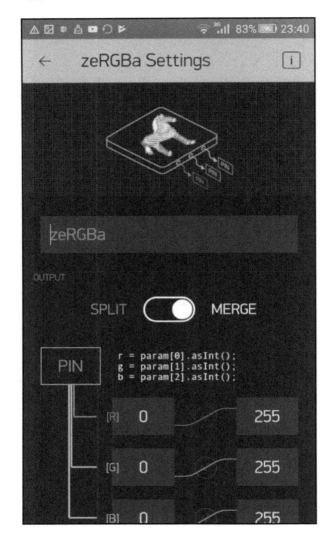

zeRGBa widget in merge mode

5. Tap **PIN**, and from the drop-down list choose **Virtual**, followed by **V1**. Then, tap **OK**.

6. The **zeRGBa Settings** page should look like this:

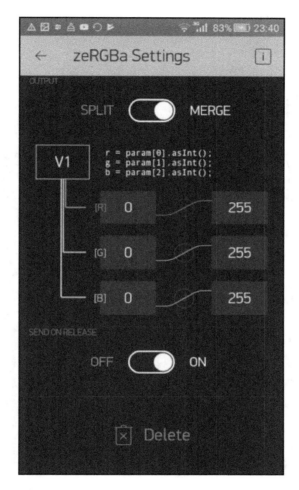

Settings for use with Virtual pins

7. Keep the other options on the settings page as they are. Tap the back arrow icon to go to the canvas. Now, your zeRGBa widget should look like this:

zeRGBa widget connected to the virtual pin V1

Controlling an RGB LED

RGB LEDs are great to demonstrate the functionality of the zeRGBa widget. The RGB LED has four pins: one for the red channel, one for the green channel, one for the blue channel, and one for ground.

The following diagram shows the pinout for a typical RGB LED with a common cathode:

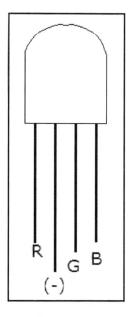

Pinout for a typical RGB LED, common cathode

Building the circuit

You will need the following things to build the circuit:

- 1 x RGB LED with common cathode (https://www.sparkfun.com/products/105)
- 3 x 220-Ohm resistors
- Small breadboard
- Hookup wires

The following diagram shows the wiring diagram between the RGB LED and the Raspberry Pi:

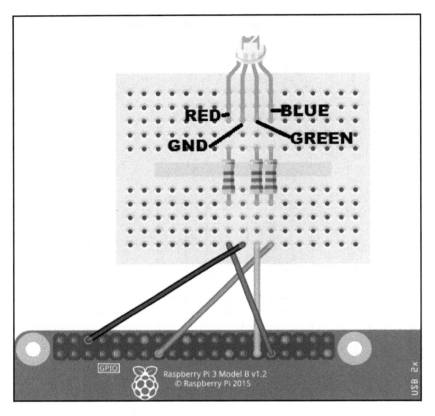

Wiring diagram for connecting an RGB LED to Raspberry Pi

Using the breadboard and the hookup wires, build the circuit:

- Connect the blue leg of the RGB LED to the Raspberry Pi BCM_GPIO Pin 22 (WiringPi Pin 3) through a 220-Ohm resistor
- Connect the green leg of the RGB LED to the Raspberry Pi BCM_GPIO Pin 5 (WiringPi Pin 21) through a 220-Ohm resistor
- Connect the red leg of the RGB LED to the Raspberry Pi BCM_GPIO Pin 6 (WiringPi Pin 22) through a 220-Ohm resistor
- Connect the GND (cathode) leg of the RGB LED to Raspberry Pi Physical Pin 6 (GND pin)

Running the project

Connect to your Raspberry Pi with your computer using PuTTY through SSH:

1. Move to the `blynk-library/linux` directory by issuing the following command:

   pi@raspberrypi:~ $`cd blynk-library/linux`

2. Open the `main.cpp` file by issuing the following command:

   pi@raspberrypi:~/blynk-library/linux $`sudonano main.cpp`

3. If not, include the software PWM library just after the WiringPi library, `wiringPi.h`:

   ```
   #include <softPwm.h>
   ```

4. Then, edit the `setup()` function as follows:

   ```
   void setup()
   {
   Blynk.begin(auth, serv, port);
   softPwmCreate(3,0,255);
   softPwmCreate(21,0,255);
   softPwmCreate(22,0,255);
   }
   ```

5. Finally, edit `BLYNK_WRITE(V1)` as follows:

   ```
   BLYNK_WRITE(V1)
   {
   printf("Got a value: %sn", param[0].asStr());
   printf("Got a value: %sn", param[1].asStr());
   printf("Got a value: %sn", param[2].asStr());
   softPwmWrite(3, param[0].asInt()); // blue channel
   softPwmWrite(21, param[1].asInt()); // green channel
   softPwmWrite(22, param[2].asInt()); // red channel
   }
   ```

6. Save the file by pressing *Ctrl + O*, followed by *Enter*, followed by *Ctrl + X*.
7. Build the C++ project by running the following command:

   pi@raspberrypi:~/blynk-library/linux $ `./build.sh raspberry`

8. After building the project, run it using the following command with the auth token associated with your Blynk app:

```
pi@raspberrypi:~/blynk-library/linux $sudo ./blynk --
token=ca7bed1c92214503a65de8e20164994f
```

9. With your Blynk app, tap the play button on the toolbar. The app will go to play mode:

zeRGBa widget in play mode

10. Tap the zebra image to select a color. When you stop touching the screen, the LED will change to the color you selected. Note that the values for red, green, and blue change according to the different selections. Any time you want, you can quickly change the color to white by tapping the white circle in the upper-left corner of the widget, and to black by tapping anywhere on the background of the widget.

Listing 3.3 shows the source code for `main.cpp`:

Listing 3.3: Controlling RGB LED with zeRGBa

```
//#define BLYNK_DEBUG

#define BLYNK_PRINT stdout

#ifdef RASPBERRY

#include <BlynkApiWiringPi.h>

#else

#include <BlynkApiLinux.h>

#endif

#include <BlynkSocket.h>

#include <BlynkOptionsParser.h>

#include <wiringPi.h>

#include <softPwm.h>

static BlynkTransportSocket _blynkTransport;

BlynkSocket Blynk(_blynkTransport);

static const char *auth, *serv;

static uint16_t port;

#include <BlynkWidgets.h>

BLYNK_WRITE(V1)

{
```

```
printf("Got a value: %sn", param[0].asStr());

printf("Got a value: %sn", param[1].asStr());

printf("Got a value: %sn", param[2].asStr());

softPwmWrite(3, param[0].asInt());

softPwmWrite(21, param[1].asInt());

softPwmWrite(22, param[2].asInt());

}

void setup()

{

Blynk.begin(auth, serv, port);

softPwmCreate(3,0,255);

softPwmCreate(21,0,255);

softPwmCreate(22,0,255);

}

void loop()

{

Blynk.run();

}

intmain(intargc, char* argv[])

{

parse_options(argc, argv, auth, serv, port);

setup();

while(true) {

loop();
```

```
}

return 0;

}
```

Joystick

The Joystick widget simulates a real joystick with two axes, x and y. You can add a Joystick widget to the canvas by following these steps:

1. Tap the plus icon to open the Widget Box.
2. Under **Controllers**, tap **JOYSTICK**. A Joystick widget will be added to the canvas.

The Joystick widget consumes 400 power:

Joystick widget

Using digital pins

In split mode, the Joystick widget uses two pins on your hardware for each axis, x and y. The pins should be capable of PWM to write values in a predefined range. Unfortunately, the Raspberry Pi has a single pin that is BCM_GPIO 18 capable with PWM, but you will require one more PWM pin. Therefore, you can't use the Joystick widget with Raspberry Pi digital pins.

Using virtual pins

These steps will explain how to use virtual pins to control the color of an RGB LED:

1. Tap **Joystick** to open the **Joystick Settings** page:

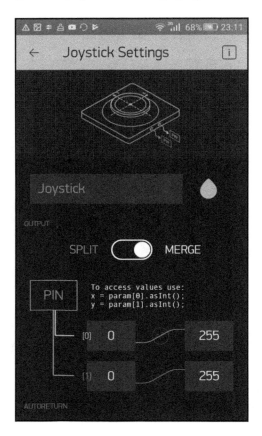

Joystick Settings

2. Tap **PIN** under **OUTPUT**. From the drop-down list, select **Virtual**, followed by **V1**. Then, tap **OK**.

3. After selecting the virtual pin V1, your **Joystick Settings** page should look like this:

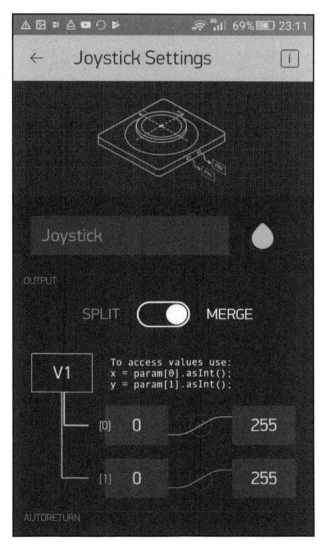

Joystick Settings for virtual pin V1

4. Don't change the other settings, such as **AUTORETURN, ROTATE ON TILT**, and **WRITE INTERVAL**:

- **AUTORETURN**: When OFF, the joystick handle will not return to the center position
- **ROTATE ON TILT**: When OFF, the joystick will automatically rotate if you use your smartphone in landscape orientation:

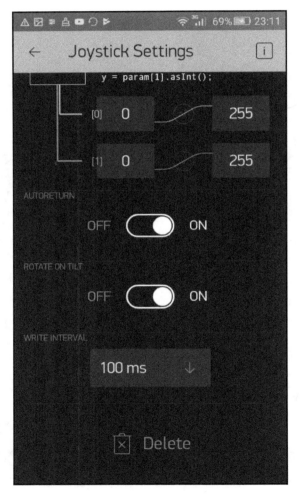

Additional settings for Joystick

5. Tap back arrow to go to the canvas. Your joystick should look as follows. The joystick writes x=**128** and y=**128** for its center position:

Joystick connected to the virtual pin V1

6. Let's modify the `main.cpp` file to print values on the console for the *x* and *y* positions of the joystick.

7. Move to the `blynk-library/linux` directory by issuing the following command:

    pi@raspberrypi:~ $cd blynk-library/linux

8. Open the `main.cpp` file by issuing the following command:

    pi@raspberrypi:~/blynk-library/linux $sudonano main.cpp

9. If not, include the software PWM library just after the WiringPi library, `wiringPi.h`:

    ```
    #include <softPwm.h>
    ```

10. Then, edit the `setup()` function as follows:

    ```
    void setup()
    {
      Blynk.begin(auth, serv, port);
    }
    ```

11. Finally, edit `BLYNK_WRITE(V1)` as shown here:

    ```
    BLYNK_WRITE(V1)
    {
    printf("x: %sn", param[0].asStr());
    printf("y: %sn", param[1].asStr());
    }
    ```

12. Save the file by pressing *Ctrl + O*, followed by *Enter*, followed by *Ctrl + X*.

13. Build the C++ project by running the following command:

    pi@raspberrypi:~/blynk-library/linux $ ./build.sh raspberry

14. After building the project, run it using the following command with the auth token associated with your Blynk app:

    pi@raspberrypi:~/blynk-library/linux $sudo ./blynk --
    token=ca7bed1c92214503a65de8e20164994f

15. With your Blynk app, tap the play button on the toolbar. The app will go to play mode:

Joystick in play mode

16. The following figure shows values for *x* and *y* when you move the joystick to the left, right, forward, and backward positions. The *x* and *y* values are always in the range 0-255:

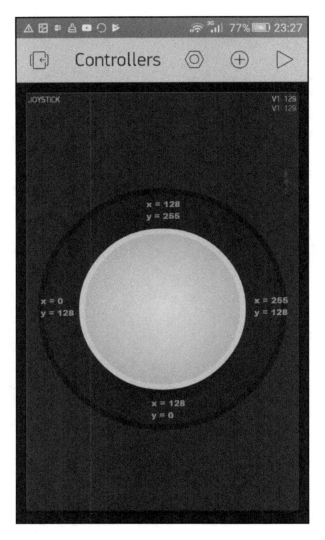

x and y values for left, right, forward, and backward positions

Listing 3.4 shows the source code for `main.cpp`:

Listing 3.4: Printing x and y values for Joystick movement

```
//#define BLYNK_DEBUG

#define BLYNK_PRINT stdout

#ifdef RASPBERRY

#include <BlynkApiWiringPi.h>

#else

#include <BlynkApiLinux.h>

#endif

#include <BlynkSocket.h>

#include <BlynkOptionsParser.h>

#include <wiringPi.h>

#include <softPwm.h>

static BlynkTransportSocket _blynkTransport;

BlynkSocket Blynk(_blynkTransport);

static const char *auth, *serv;

static uint16_t port;

#include <BlynkWidgets.h>

BLYNK_WRITE(V1)

{

printf("x: %sn", param[0].asStr());

printf("y: %sn", param[1].asStr());

}

void setup()
```

```
{

Blynk.begin(auth, serv, port);

}

void loop()

{

Blynk.run();

}

int main(intargc, char* argv[])

{

parse_options(argc, argv, auth, serv, port);

setup();

while(true) {

loop();

}

return 0;

}
```

Summary

In this chapter, you learned how to use controller widgets such as Slider, Step, zeRGBa, and Joystick with your Blynk app to control actuators with digital and virtual pins. Using virtual pins allows you to overcome the PWM pin limitation of the Raspberry Pi, as the Raspberry Pi has only one PWM pin, BCM_GPIO 18. You also used the software PWM library, which is the part of the WiringPi library used to define any GPIO pin as a PWM-enabled pin to work with controllers that have multiple outputs in merge mode.

In Chapter 4, *Using Display Widgets*, you will learn how to build applications with display widgets.

4

Using Display Widgets

Display widgets allow you to display any data coming from the hardware. As an example, you can display temperature readings on the Blynk app that come from a temperature sensor attached to the Raspberry Pi. The Widget Box provides different types of display widgets, so you can choose one based on your display requirement. In this chapter, you will learn about some of the important display widgets that you can use with the Blynk app builder:

- Value Display
- Labeled Value
- LED

Before adding any display widget to your canvas, delete any existing controller widgets to save energy, because each widget costs energy. You can also rename the project to displays. However, the screen captures presented in this chapter used the same Blynk application, named Controllers, without renaming it.

Value Display

The Value Display widget can be used to display incoming data from the sensors attached to the Raspberry Pi. You can use one of the following mechanisms to display data on the Value Display widget:

- **Perform requests by the widget**: The widget in the app reads a pin with a certain frequency. This is also called a pull model
- **Pushing data from the hardware**: Send data from the hardware to the app widget at intervals. This is also called a push model

You can add a Value Display widget onto your Blynk app with the **Widget Box** by tapping **Value Display** under **DISPLAY**. The Value Display widget costs 200 units of energy.

The default label is **VALUE**, as shown in the following screenshot:

Value Display widget added to the canvas

Using digital pins

Let's build a Blynk app and write code to display data coming from the potentiometer attached to the Raspberry Pi using digital pins. To build the project, you will need the following things:

- A potentiometer
- Breadboard
- Hookup wires

Following diagram shows the wiring diagram for building the circuit. Connect the center pin of the potentiometer to the Raspberry Pi BCM_GPIO pin 18 (WiringPi pin 1). Connect one of the outer pins of the potentiometer with the Raspberry Pi 5V pin. Connect the other outer pin of the potentiometer with Raspberry Pi GND:

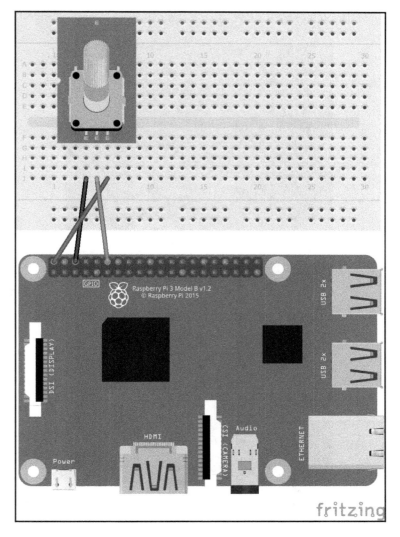

Potentiometer attached to the Raspberry Pi BCM_GPIO pin 18 (WiringPi pin 1)

Now, let's configure the Value Display widget with the digital pin (BCM_GPIO pin 18 / WiringPi pin 1):

1. Tap the **Value Display** widget to open the **Value Display Settings** page.
2. Tap **PIN**, and from the left-hand side list, select **Digital**. Then, from the right-hand side list, select **gp 18 PWM** (see the following screenshot):

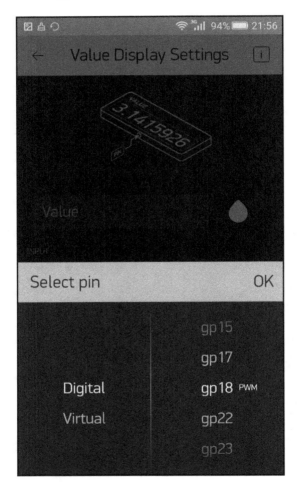

Selecting a digital pin

3. Tap **OK** and now your **Value Display Settings** page should look like this (see the following screenshot):

Value Display widget configured with a digital pin

4. Tap the back button on the toolbar to view the canvas. The Value Display widget is now configured with digital pin BCM_GPIO pin 18 (WiringPi pin 1), as shown in the following screenshot:

Configured Value Display widget on canvas

5. Modify the `main.cpp` file as shown in *Listing 4.1* (`https://github.com/ PacktPublishing/Hands-On-Internet-of-Things-with-Blynk/blob/master/ Chapter%204/Listing%204-1/main.cpp`). Use the following command to locate and open the file with the PuTTY terminal by connecting through SSH to the Raspberry Pi:

```
pi@raspberrypi:~ $ cd blynk-library/linux
pi@raspberrypi:~/blynk-library/linux $sudonano main.cpp
```

6. Save the file and exit the editor by pressing *Ctrl + O*, followed by *Enter*, followed by *Ctrl + X*.

7. Now, build the C++ project with the following command:

```
pi@raspberrypi:~/blynk-library/linux $ ./build.sh raspberry
```

8. Run the C++ project using the following command with the auth token associated with your Blynk project:

```
pi@raspberrypi:~/blynk-library/linux $sudo ./blynk --
token=ca7bed1c92214503a65de8e20164994f
```

9. Now, tap the play button to start the app. The Value Display widget will show the current analog reading value of the potentiometer. You can take potentiometer values between 0 and 1023.

Using virtual pins

The Value Display widget also supports virtual pins. Follow these steps to configure the Value Display widget with a virtual pin:

1. Tap the stop button to stop the application and go back to the canvas.
2. Tap the Value Display widget to open the **Value Display Settings** page.
3. Under **INPUT**, tap **gpio 18 PWM**. From the left-hand side list, select **Virtual**, and from the right-hand side list, select **V1**.

4. Tap **OK**. Now, your **Value Display Settings** page should look like the following (see the following screenshot):

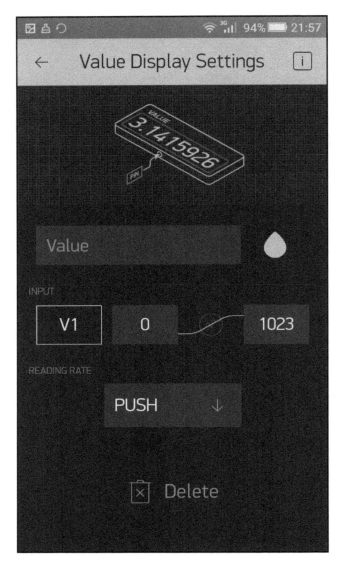

Value Display settings for a virtual pin

5. Tap the back button on the toolbar to go to the canvas page. Now, you can see the **Value Display widget** is configured with the virtual pin **V1**.

6. Modify the `main.cpp` file as shown in *Listing 4.2* (`https://github.com/PacktPublishing/Hands-On-Internet-of-Things-with-Blynk/blob/master/Chapter%204/Listing%204-2/main.cpp`). Use the commands in the previous section to locate and open the file.

7. Save and exit the editor by pressing *Ctrl + O*, followed by *Enter*, followed by *Ctrl + X*.

8. Build the C++ project using this command:

```
pi@raspberrypi:~/blynk-library/linux $ ./build.sh raspberry
```

9. Run the C++ project using this command, with the auth token associated with your Blynk application:

```
pi@raspberrypi:~/blynk-library/linux $sudo ./blynk --
token=ca7bed1c92214503a65de8e20164994f
```

10. Now, tap the play button to start the app. The Value Display widget will show the analog reading value of the potentiometer. You can take potentiometer values between 0 and 1023.

Labeled Value

The Labeled Value widget is very similar to the Value Display widget, but allows you to format the output the way you want. As an example, you can append the text C to the sensor values coming from a temperature sensor with the Labeled Value widget.

The Labeled Value widget supports both digital and virtual pins. You can add a Labeled Value widget to the Blynk app by tapping Labeled Value under **DISPLAYS** in the **Widget Box**.

After adding a Labeled Value widget, your canvas should look like this (see the following screenshot):

Labeled Value widget

Configuring a Labeled Value widget

You can get the configuration settings page for the Labeled Value widget by tapping the Labeled Value widget on the canvas. Following screenshot shows the **Labeled Value Settings** page:

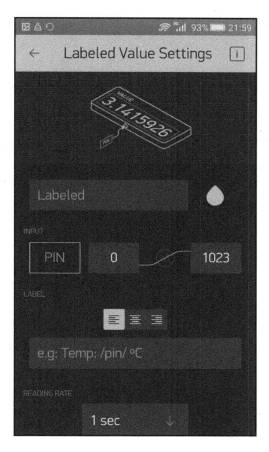

Labeled Value Settings page

The configuration for the digital and virtual pins are the same as the Value Display widget discussed in the previous section. The additional feature, **LABEL**, can be used for formatting the output display on the label.

Under LABEL, tap the textbox to edit. The formatting options are listed here. For example, your sensor sends a value of `12.6789` to the Blynk application, and you can format it using one of the following options:

- `/pin/`: Displays the value without formatting (`12.6789`)
- `/pin./`: Displays the rounded value without the decimal part (`13`)
- `/pin.#/`: Displays the value to one decimal place (`12.7`)
- `/pin.##/`: Displays the value to two decimal places (`12.68`)

You can also change the alignment of the output using one of the three buttons: left, center, and right.

The sample code you used in the Value Display widget can be used with the Labeled Value widget without making any modifications.

LED

The LED widget simulates an LED and it can be used as an indicator to present the status or amount of an input. As an example, you can use it to present the status of a switch or output value of a potentiometer attached to the Raspberry Pi.

You can add an LED widget onto your Blynk app with the **Widget Box** by tapping **LED** under **DISPLAYS**. The LED widget costs **100** units of energy (see the following screenshot):

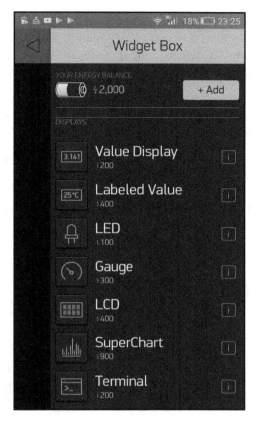

LED widget listed under DISPLAYS

Using virtual pins

The LED widget supports only virtual pins. Using a momentary push button with Raspberry Pi is a great way to learn how virtual pins work with inputs. In this example, you will see how the LED widget can be turned on and off according to the status of the momentary push button switch. Let's build the circuit using the following items:

- Momentary push button
- Breadboard
- Hookup wires

Build the circuit as shown in the following screenshot. Connect the push button between Raspberry Pi BCM_GPIO pin 23 (WiringPi pin 4) and the GND. If you want, you can attach the push button to the WiringPi pin 1 (BCM_GPIO 18), the only pin that supports PWM output, and modify the code accordingly:

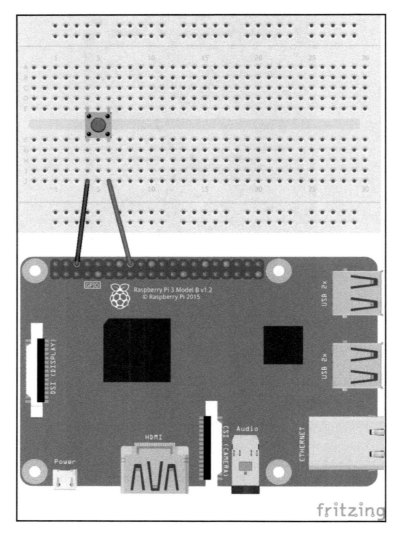

Push button connected to the BCM_GPIO pin 23 (WiringPi pin 4)

The **LED Settings** page allows you to connect the LED widget to an **INPUT** pin, digital or virtual:

1. Tap the LED widget to open the **LED Settings** page, which will look like this (see the following screenshot):

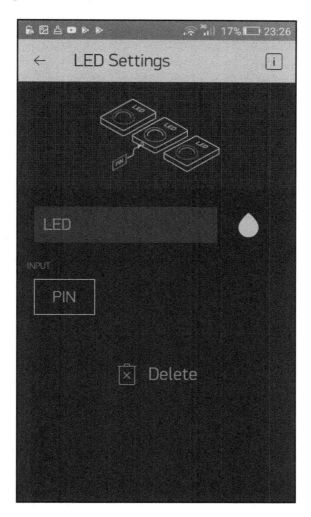

LED Settings page

2. Tap **PIN** under **INPUT**.

3. Select **Virtual** from the left-hand side list and **V1** from the right-hand side list. Then, tap **OK**.

4. Tap the back arrow on the toolbar to go to the canvas. The LED widget is now configured with virtual pin **V1**.

5. Connect to the Raspberry Pi using PuTTY through SSH. Then, open the `main.cpp` file using these commands:

```
pi@raspberrypi:~ $ cd blynk-library/linux
pi@raspberrypi:~/blynk-library/linux $sudonano main.cpp
```

6. Modify the `main.cpp` file as shown in *Listing 4.3* (`https://github.com/PacktPublishing/Hands-On-Internet-of-Things-with-Blynk/blob/master/Chapter%204/Listing%204-3/main.cpp`).

7. Build the project by issuing the following command:

```
pi@raspberrypi:~/blynk-library/linux $ ./build.sh raspberry
```

8. Run the project with the auth token associated with the project:

```
pi@raspberrypi:~/blynk-library/linux $sudo ./blynk --
token=ca7bed1c92214503a65de8e20164994f
```

9. Tap the play button on the Blynk app toolbar. The app will connect to the Blynk cloud and is ready to receive the button states coming from the Raspberry Pi through the virtual pin **V1**.

10. Now you can turn the LED widget on and off with the momentary push button attached to the Raspberry Pi.

Summary

In this chapter, you learned how to use some important display widgets with your Blynk projects. Display widgets help you to display any data on the Blynk app that comes from the hardware. Most of the display widgets support both digital and virtual pins, but some only support Virtual pins. In Chapter 5, *Using Notification Widgets*, we will focus on accessing and using third-party services for Blynk from Raspberry Pi.

Using Notification Widgets

5

Notification widgets are another subset of widgets that can be used to send notifications from your Raspberry Pi hardware. Some notification widgets can be integrated with third-party services such as Twitter and Blynk cloud. You can schedule your Raspberry Pi to send notifications to your smartphone (or tablet) or you can send them on a user action (if a button is pressed or depending on a threshold value).

In this chapter, you will learn:

- How to use the Twitter widget to send tweets from your Raspberry Pi with your Twitter account
- How to use the notification widget to send pop-up notifications to a smartphone or tablet
- How to use the Email widget to send emails through the Blynk cloud (`https://www.blynk.io/`)

Twitter

The Twitter widget allows you to connect your Twitter account with Blynk and send tweets from your Raspberry Pi hardware. However, the Twitter widget has the following limitations:

- You can't tweet the same message more than once
- Only one tweet per 15 seconds is allowed

Adding a Twitter widget

These steps will show you how to add a Twitter widget onto your canvas:

1. In the **Widget Box**, tap Twitter under **NOTIFICATIONS**:

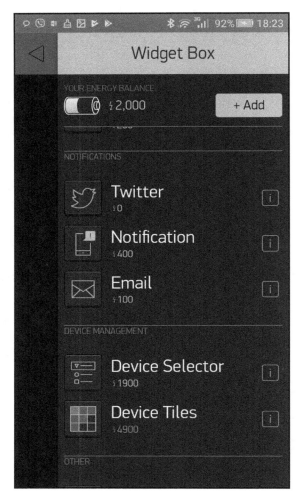

Twitter widget

2. The Twitter widget will be added to the canvas. However, you can't add more than one Twitter widget to a project. A single Twitter widget can handle all the tweets sent from your Raspberry Pi. The Twitter widget consumes 0 units of energy from your energy bank:

Twitter widget added to the canvas

Configuring

You should provide your Twitter account's login credentials to connect the Twitter widget with Twitter. OAuth allows Blynk to tweet from your Twitter account without actually sharing any credentials (no need to mention login credentials in the code). After connecting, it will use your Twitter account to send tweets. The tweets appear just as if you sent them from the Twitter website.

Currently, the Twitter widget only supports sending tweets from your Twitter account. It doesn't support retweeting, so you can't use your Raspberry Pi as a retweet bot:

1. Tap the Twitter widget to open the **Twitter Settings** page.
2. In the **Twitter Settings** page, tap the **Connect Twitter** button:

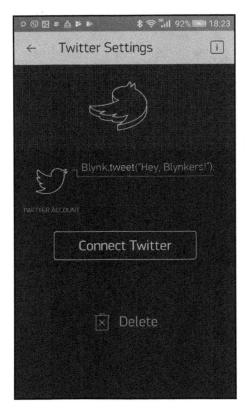

Twitter Settings page

3. The **Sign up for Twitter** page will appear. You can use this page to authorize the Blynk to use your Twitter account.

4. Type in your Twitter account's username (or email address) and password, then tap the **Authorise app** button to connect with your Twitter account:

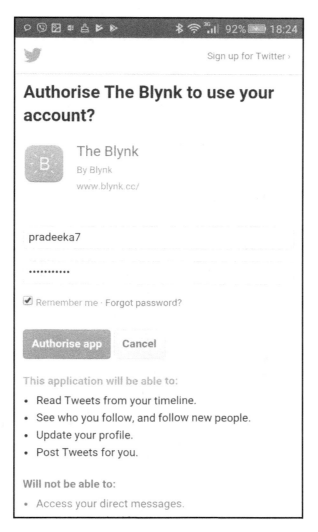

Authorization page to use the Twitter account

5. After a few seconds the **Twitter Settings** page will show you **connected**, along with your Twitter account name, if the authorization is successful. You can disconnect your Twitter account from Blynk by tapping the **Disconnect** button:

Authorization status

6. Now, tap the back arrow on the toolbar to go to the canvas.
7. On the canvas page, on the toolbar, tap the play button to start the app.

Sending tweets

The Blynk library provides the `Blynk.tweet()` function to send tweets of up to 128 characters:

```
Blynk.tweet("Tweeting from Raspberry Pi");
```

However, you can increase the message length by including the following statement at the top of the code:

```
#define BLYNK_MAX_SENDBYTES 256
```

The example code shown in *Listing 5.1* shows how to send a tweet from the Raspberry Pi every 30 seconds:

Listing 5.1: Send a tweet from Raspberry Pi in every 30 seconds

Refer to the following steps:

1. Open the `main.cpp` file and type in the preceding code.
2. The `SimpleTimer` library is used to handle timed actions. You can instantiate the `SimpleTimer` library to create an object of the `SimpleTimer`:

   ```
   SimpleTimer timer;
   ```

3. The `SimpleTimer` library is based on `millis()`; therefore it has 1 ms resolution.
4. The `setInterval()` function of the `SimpleTimer` library allows you to call a function with specified intervals in milliseconds. The preceding code calls the `tweetMe()` function every 30 seconds. It accepts the time in milliseconds; therefore, 30 seconds is equivalent to `30*1000` milliseconds. You can change the intervals between tweets by adjusting the time, but don't use less than 15 seconds:

   ```
   void tweetMe() {
   ...
   }
   timer.setInterval((30*1000), tweetMe);
   ```

5. The `uptime` variable holds the time in milliseconds since the program started on the Raspberry Pi:

```
intuptime = millis()/1000;
```

6. A unique tweet can be built by concatenating a constant string with the `uptime` in seconds as follows. The `sprintf()` function is used to convert the `uptime` (an integer) into a string. The `strcat()` function is then used to concatenate the two strings to build a single string. Make sure to build the string under 128 bytes unless you defined `#define BLYNK_MAX_SENDBYTES 256` at the top of the code to use 255 bytes:

```
char msg[] = "My Raspberry Pi is tweeting using @blynk_app and
it's awesome!nTweeting time since startup: ";
char tm[100];
sprintf(tm,"%d",uptime);
strcat(msg,tm);
strcat(msg," seconds.");
Blynk.tweet(msg);
```

7. Save the file and exit from the nano text editor by pressing *Ctrl + O*, followed by *Enter,* followed by *Ctrl + X.*

8. Build the C++ project with the following command:

pi@raspberrypi:~/blynk-library/linux $ ./build.sh raspberry

9. Now, run your C++ project with the associated auth token from your computer:

**pi@raspberrypi:~/blynk-library/linux $sudo ./blynk --
token=ca7bed1c92214503a65de8e20164994f**

10. Log in to your Twitter account and go to the **MyTweets** page. Wait for 30 seconds, and you will see the first tweet on the top of the page. The next tweet will appear 60 seconds after starting your program, and so on:

Tweeting from Raspberry Pi every 30 seconds

Twitter button

Let's build another project based on the Twitter widget, so you can send tweets for every button press event.

You will need the following things to build the project:

- Raspberry Pi
- Momentary push button
- Breadboard
- Hookup wires

The following diagram shows the wiring diagram that you can use to build the circuit with the Raspberry Pi:

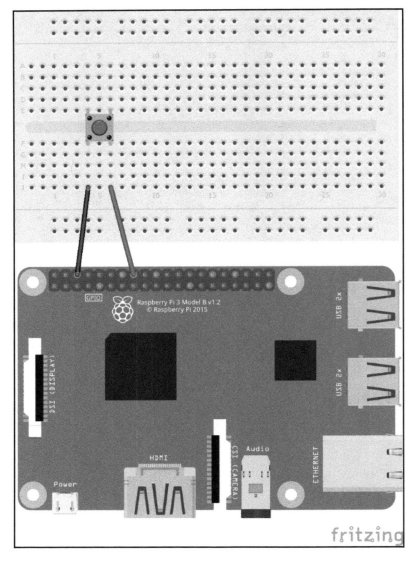

Push button is connected between BCM_GPIO pin 23 and 0V

The push button is connected between the BCM_GPIO pin 23 (physical pin 16) and the 0V (physical pin 6).

Listing 5.2 shows the C++ source code for detecting the button press event and then sending a tweet:

Listing 5.2: Send a tweet when the push button is pressed

The code is very similar to the code listed in *Listing 5.1,* but in addition, it includes code to detect the button press:

1. The push button is connected to the BCM_GPIO pin 23. First, you should configure pin 23 as an input pin:

   ```
   pinMode(btnPin, INPUT);
   ```

2. By default, the BCM_GPIO pin is set to pull-up mode with the following statement:

   ```
   pullUpDnControl(btnPin, PUD_UP);
   ```

3. With pull-up mode, the value of pin 23 is digitally HIGH. When you press the button, pin 23 will connect to the 0V and go to pull-down mode. Now, pin 23 presents digital low until you release the button.
4. Press *Ctrl + O*, followed by *Enter*, followed by *Ctrl + X* to save and exit from the nano text editor.
5. Build the project with the command shown here:

   ```
   pi@raspberrypi:~/blynk-library/linux $ ./build.sh raspberry
   ```

6. Run the project with the auth token:

   ```
   pi@raspberrypi:~/blynk-library/linux $sudo ./blynk --
   token=ca7bed1c92214503a65de8e20164994f
   ```

7. Now, open your Twitter page and go to Tweets. Then, press and release the momentary push button. Immediately, you will be notified with a new tweet that came from the Raspberry Pi, tweeted under your account name:

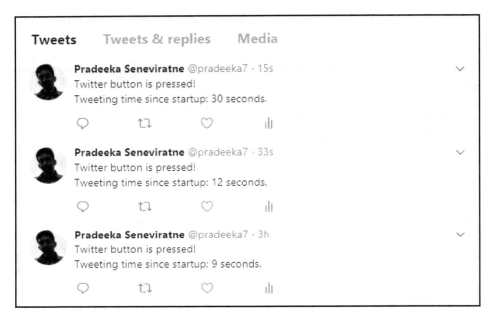

Sending tweets when the button is pressed

Remember, you can send one tweet per 15 seconds. It also doesn't allow you to send identical subsequent messages. To make a message unique, the uptime since the program started is added to the message.

Notification widget

The Notification widget allows you to send push notifications from your Raspberry Pi to your smartphone or tablet.

This has the following limitations:

- The maximum message length is 128 characters (mix of alphanumeric and symbols)
- You can send only one notification per 15 seconds

However, you can increase the message length by including the following statement at the beginning of the code:

```
#define BLYNK_MAX_SENDBYTES 256 // Default is 128
```

You can add a Notification widget onto your project canvas using the Widget Box. The Notification widget also consumes 0 units of energy from your energy bank. Only one notification widget is allowed for a project:

Notification widget is added next to the Twitter widget

Configuring the Notification widget

Refer to the following steps:

1. Tap the Notification widget to get the **Notification Settings** page:

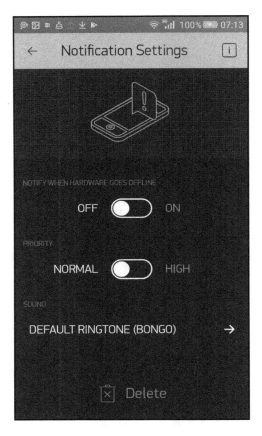

Notification Settings page

2. The Notification widget settings provide the following options:

- **NOTIFY WHEN HARDWARE GOES OFFLINE**: The user will get a push notification to their smartphone or tablet if the Raspberry Pi goes offline. Under **NOTIFY WHEN HARDWARE GOES OFFLINE**, simply slide the button to the ON position to enable this feature.
- **OFFLINE IGNORE PERIOD**: If you enable this option, you can further define how long the Raspberry Pi can be offline before sending a notification:

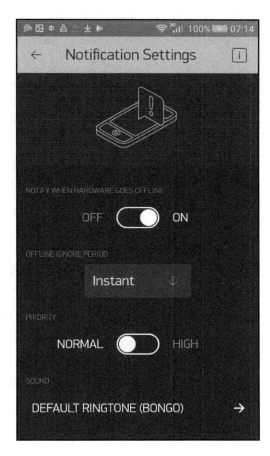

Configuring offline ignore period

3. Under **OFFLINE IGNORE PERIOD**, tap the drop-down list and choose one of the items. **Instant** means there is no ignore period, so the Raspberry Pi will immediately send a notification to the user if it goes offline. You can choose the time, between **1** second and 24 hours:

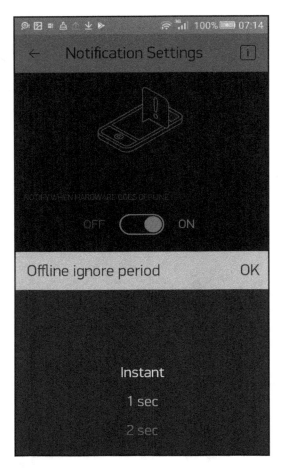

Choosing offline ignore period

- **PRIORITY**: The second option, **PRIORITY**, could be either **NORMAL** or high. **HIGH** priority gives a greater chance that your message will be delivered without any delays.
- **RINGTONE**: The last option allows you to select a ringtone for the notification. The default ringtone is **BONGO!** You can select a ringtone from **Media Storage** or **Themes** from your smartphone or tablet:

Selecting a ringtone from media storage or themes

Blynk allows you to send one notification every 15 seconds.

Writing a notification

Let's build a simple project to send a notification to the smartphone or tablet for the button press event. You can use the same hardware setup you built in the previous section:

Listing 5.3 shows the complete code to send the notification from your Raspberry Pi:

Listing 5.3: sending notification for the button press event

The Blynk.notify() function allows you to mention your own message as a notification:

```
Blynk.notify("Yaaay... button is pressed!");
```

Refer to the following steps:

1. Type in the preceding code by opening the main.cpp file. Then, save and exit the nano text editor by pressing *Ctrl + O*, followed by *Enter*, followed by *Ctrl + X*.
2. After that, build the project by issuing the following command:

   ```
   pi@raspberrypi:~/blynk-library/linux $ ./build.sh raspberry
   ```

3. Now, you can run the project with the auth token associated with the project:

   ```
   pi@raspberrypi:~/blynk-library/linux $sudo ./blynk --
   token=ca7bed1c92214503a65de8e20164994f
   ```

4. To test the application, press and release the push button attached to BCM_GPIO pin 23 of the Raspberry Pi. Immediately, you will get a notification on your smartphone or tablet:

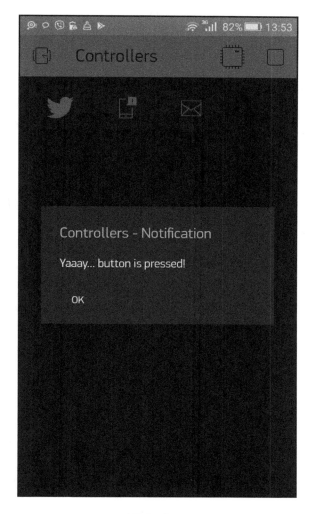

A Blynk notification

Email

Email is another useful widget that allows you to send emails from your Raspberry Pi through the Blynk server with the Raspberry Pi hardware.

You can send an email with the Blynk cloud under the following conditions:

You can send only one email per 15 seconds, but you will also be limited to only 100 emails per day, You can add the Email widget onto your project canvas using the Widget Box as usual:

Email widget placed next to the notification widget

Configuring the Email widget

Refer to the following steps:

1. To configure the Email widget, tap on it to get the **Email Settings** page.

2. In the **Email Settings** page, under **TO**, type in your recipient's email address.

3. You can choose the content type as either `text/html` or `text/plain`:

Email Settings page

4. Tap the back arrow to go to the project canvas.

Writing code to send email

Listing 5.4 shows sample code that you can use to send emails from the Raspberry Pi when a button is pressed. You can use the same hardware setup you used in the previous section to build this application:

1. The push button is currently attached to the BCM_GPIO pin 23:

    Listing 5.4: Sending Emails from Raspberry Pi when the push button is pressed

    ```
    #define BLYNK_PRINT stdout

    #ifdef RASPBERRY

    #include <BlynkApiWiringPi.h>

    #else

    #include <BlynkApiLinux.h>

    #endif

    #include <BlynkSocket.h>

    #include <BlynkOptionsParser.h>

    #include <wiringPi.h>

    #include <softPwm.h>

    static BlynkTransportSocket _blynkTransport;

    BlynkSocket Blynk(_blynkTransport);

    #include <stdlib.h>

    #include <BlynkWidgets.h>

    SimpleTimer timer;

    constintbtnPin = 23;

    void notifyMe()

    {
    ```

```
if(digitalRead(btnPin)== LOW)

{

Blynk.email("recipient@example.com", "Hello", "Hello
Blynkers");

}

}

void setup()

{

wiringPiSetup();

pinMode(btnPin, INPUT);

pullUpDnControl(btnPin, PUD_UP);

timer.setInterval(1000, notifyMe);

}

void loop()

{

Blynk.run();

timer.run();

}

int main(intargc, char* argv[])

{

const char *auth, *serv;

uint16_t port;

parse_options(argc, argv, auth, serv, port);

Blynk.begin(auth, serv, port);

setup();
```

```
while(true) {

loop();

}

return 0;

}
```

2. The `Blynk.email()` function allows you to send a maximum of 120 symbols including email, subject, and message:

```
Blynk.email("recipient@example.com", "Hello", "Hello
Blynkers");
```

3. However, you can increase this limit if necessary by adding **#define BLYNK_MAX_SENDBYTES N** at the beginning of the code, where N can have a maximum value of `1200`. As an example, if N = `600`, the statement should be:

```
#define BLYNK_MAX_SENDBYTES 600
```

4. Type in the preceding code by opening the `main.cpp` file. Then, save and exit from the nano text editor by pressing *Ctrl + O*, followed by *Enter*, followed by *Ctrl + X*.

5. After that, build the project by issuing the following command:

```
pi@raspberrypi:~/blynk-library/linux $ ./build.sh raspberry
```

6. Now, you can run the project with the auth token associated with the project:

```
pi@raspberrypi:~/blynk-library/linux $sudo ./blynk --
token=ca7bed1c92214503a65de8e20164994f
```

7. To test the application, press and release the push button attached to the BCM_GPIO pin 23 of the Raspberry Pi. Immediately, you will get an email; if you can't find the email in your inbox, then you can check in the spam or junk folder of your email client application:

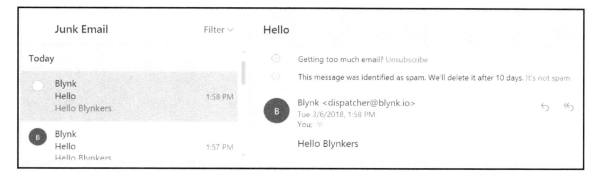

Email found in junk folder

Summary

In this chapter, you learned all about the notification widgets included with the Blynk app builder. You built various applications to send notifications such as tweets, pop-up notifications, and emails based on user actions.

In Chapter 6, *Connecting with Sensors on Your Mobile Device*, you will learn how to get data from built-in sensors on your mobile device (Android/iOS) and display it in the Blynk application.

6
Connecting with Sensors on Your Mobile Device

Modern smartphones and tablets are equipped with lots of sensors to sense their environment. These sensors can be easily integrated with your Blynk applications. In this chapter, you will read data from the following built-in sensors on your smartphone or tablet:

- Accelerometer
- Light sensor
- Proximity sensor

Accelerometer

Most smartphones and tablets come with a built-in accelerometer. An accelerometer is an electromechanical device that measures acceleration, which is the rate of change of the velocity of the phone. It measures in meters per second squared (m/s^2) or in G-forces (g). An accelerometer can sense either static or dynamic forces of acceleration:

- Sense static forces, including gravity
- Sense dynamic forces, including vibrations and movement

For your smartphone or tablet, accelerometers are useful for sensing vibrations in systems or for the orientation of applications.

The accelerometer on your smartphone measures linear acceleration along x, y, and z axes. *Figure 6.1* shows the x, y, and z orientation axes relative to a typical Android mobile device:

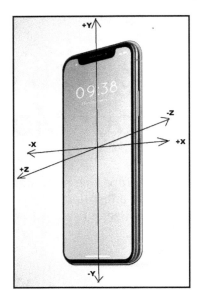

Figure 6.1: x, y, and z orientation axes for smartphone (created by Freepik)

If you place the smartphone face-up on a table, the z-axis measures the acceleration of Earth's gravity and outputs ≈ 9.81 in m/s^2. The x and y axes, which are perpendicular to the acceleration of the Earth's gravity, both output ≈ 0.00 in m/s^2.

Accelerometer widget

The Accelerometer widget allows you to read accelerometer data from your mobile device. In other words, it can internally connect with the built-in accelerometer on your mobile device and read data.

These steps will show you how to add the Accelerometer widget onto the project's canvas:

1. On the toolbar, tap the plus icon to open the Widget Box.
2. Then, scroll down the page, and under **SMARTPHONE SENSORS**, tap **Accelerometer**. The Accelerometer widget takes 400 units of energy from your energy store.
3. The Accelerometer widget will be placed on the canvas (*Figure 6.2*):

Figure 6.2: Accelerometer widget is placed on the canvas

 You can only have a single Accelerometer widget in your project.

Configuring the Accelerometer widget

The Accelerometer widget only works with virtual pins. The widget takes readings from the built-in accelerometer in your smartphone and sends them as an array to the Raspberry Pi. This mode is known as merge mode. You can access the elements of the received array to get the accelerometer readings for three axes:

```
BLYNK_WRITE(V1) {

//acceleration force applied to axis x

int x = param[0].asFloat();

//acceleration force applied to axis y

int y = param[1].asFloat();

//acceleration force applied to axis y

int z = param[2].asFloat();

}
```

These steps will show you how to configure the Accelerometer widget with virtual pin V1:

1. In the canvas view, tap the Accelerometer widget. You will get the **Accelerometer Settings** page.
2. Under **OUTPUT**, tap **PIN**. In the **Select pin list**, scroll down and choose **Virtual**, followed by **V1**, and tap **OK**.

3. Under **WRITE INTERVAL**, from the drop-down list, choose **100 ms** (*Figure 6.3*):

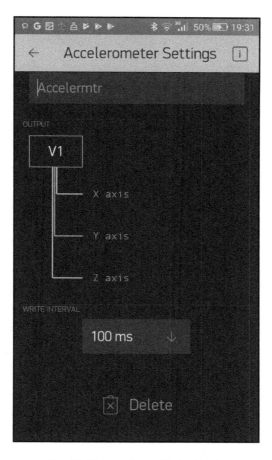

Figure 6.3: Selecting a write interval for the virtual pin

4. After you have done all the configuration, tap the back arrow on the toolbar to go to the canvas. Now, your Accelerometer widget should look like this (*Figure 6.4*):

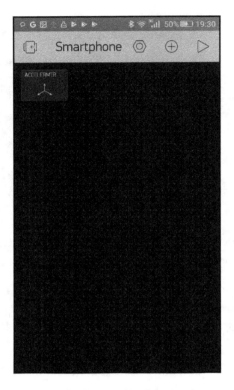

Figure 6.4: Accelerometer widget after configuration

Reading accelerometer data

Now you're ready to write a code on your Raspberry Pi to read the accelerometer on your smartphone. The code will output accelerometer values on the x, y, and z axes of the smartphone or tablet in m/s^2:

1. Modify the `main.cc` file with the source code listed in *Listing 6.1*:

   Listing 6.1: Reading the accelerometer on the smartphone

2. Then, save the file and exit the nano text editor by pressing *Ctrl + O*, followed by *Enter*, followed by *Ctrl +X*.

3. Build the project with the following command:

```
pi@raspberrypi:~/blynk-library/linux $ ./build.sh raspberry
```

4. Then, run the program with the auth token associated with your Blynk project:

```
pi@raspberrypi:~/blynk-library/linux $sudo ./blynk --
token=ca7bed1c92214503a65de8e20164994f
```

5. Tap the play button on the Blynk app builder to start the Blynk app on your mobile device. The console will output the readings for accelerometer values on the *x*, *y*, and *z* axes.

6. Let's take some accelerometer values for different gestures, such as face-up, up, and left.

7. Place your mobile device face-up on a table, as shown in *Figure 6.5*; the *z*-axis measures the acceleration of Earth's gravity and outputs ≈9.81 in m/s^2. The *x* and *y* axes, which are perpendicular to the acceleration of Earth's gravity, both output ≈0.00 in m/s^2:

Figure 6.5: Face-up gesture (created by Zlatko_plamenov—Freepik.com)

8. The following screenshot shows the accelerometer readings for the face-up gesture:

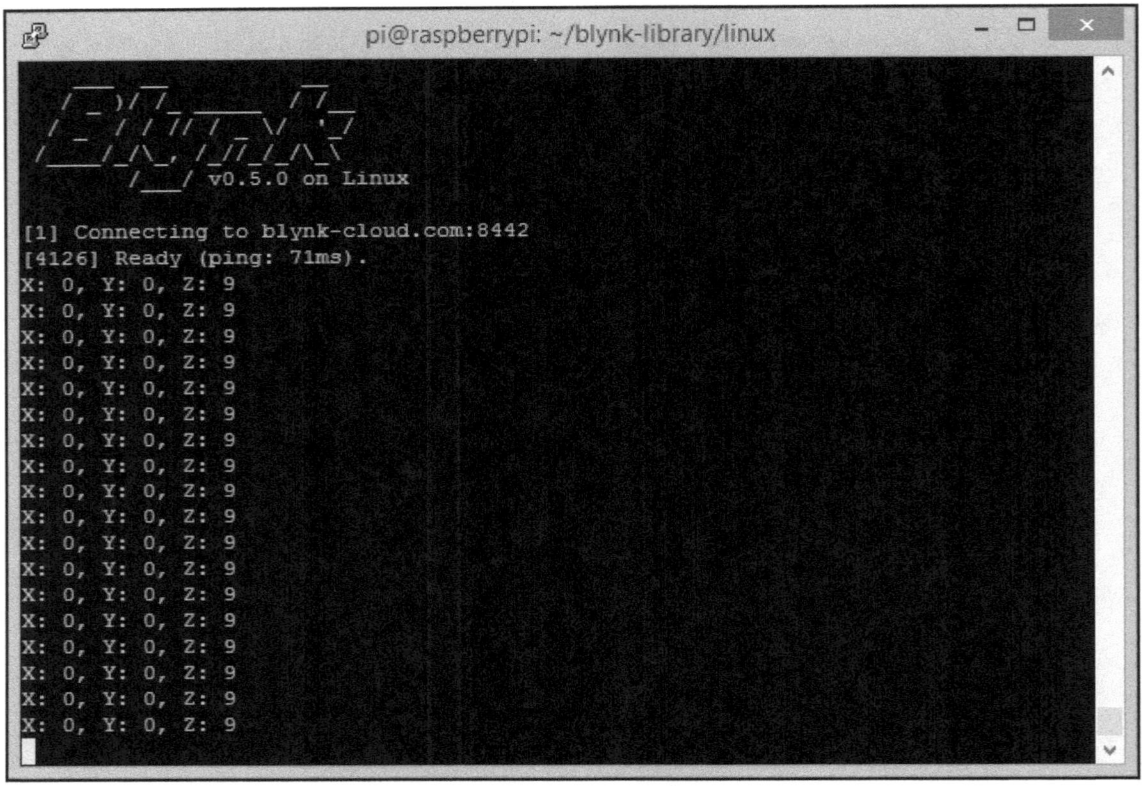

Figure 6.6: Accelerometer output for face-up gesture

9. For the opposite gesture of face-up, which is face-down, the accelerometer will produce negative values on the z-axis (*Figure 6.7*):

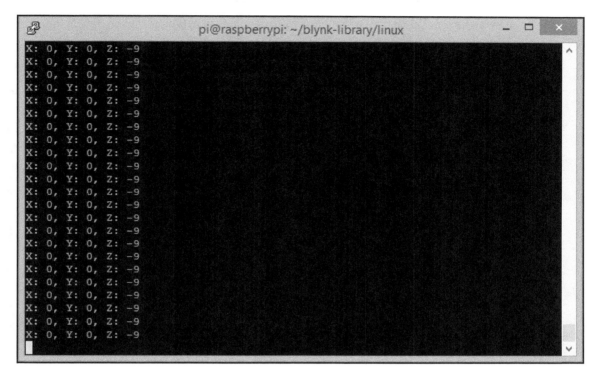

Figure 6.7: Accelerometer output for face-down gesture

10. Then, place your mobile device on a table (so the mouthpiece side is on the table), as shown in *Figure 6.8*. The *y*-axis measures the acceleration of Earth's gravity and outputs ≈9.81 in m/s². The *x* and *z* axes, which are perpendicular to the acceleration of Earth's gravity, both output ≈0.00 in m/s²:

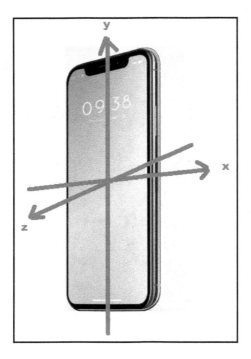

Figure 6.8: Up gesture (mouthpiece side is on the table)—Created by Freepik

11. The following screenshot shows the output values on each axis for this gesture:

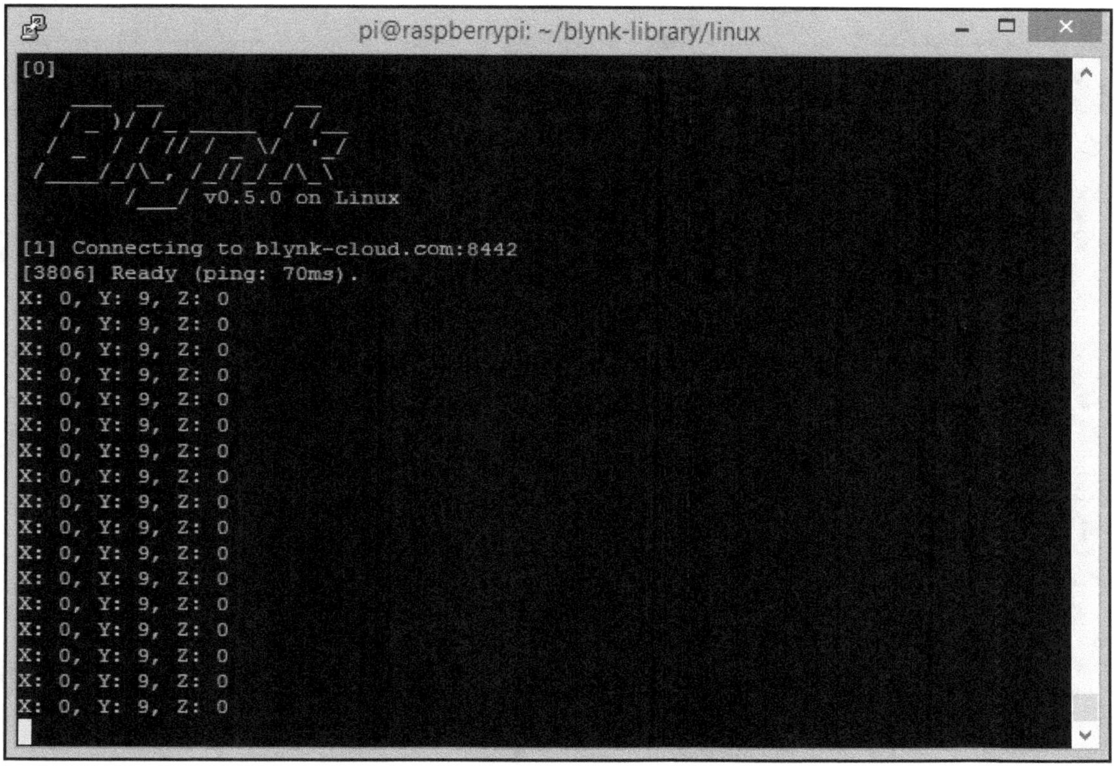

Figure 6.9: Accelerometer output for up gesture

12. If you place the earpiece side of the mobile device on the table (down gesture), the accelerometer will produce negative values on the *y* axis (*Figure 6.10*):

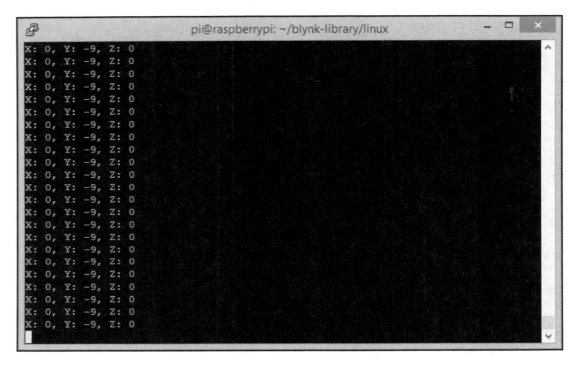

Figure 6.10: Accelerometer output for down gesture

13. Finally, place your mobile device on a table (the left-hand side of the mobile device should be on the table), as shown in *Figure 6.11*. The *x*-axis measures the acceleration of Earth's gravity and outputs ≈9.81 in m/s². The *y* and *z* axes, which are perpendicular to the acceleration of Earth's gravity, both output ≈0.00 in m/s²:

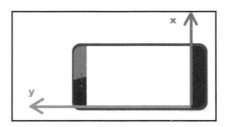

Figure 6.11: Left gesture (created by Freepik)

14. *Figure 6.12* shows the output values on each axis for this gesture:

```
pi@raspberrypi: ~/blynk-library/linux

/  _  / / / // /   \/  '_/
/___/ /\ , / // / /\ \
     /___/ v0.5.0 on Linux

[2] Connecting to blynk-cloud.com:8442
[3262] Ready (ping: 71ms).
X:  9,  Y:  0,  Z:  0
X:  9,  Y:  0,  Z:  0
X:  9,  Y:  0,  Z:  0
X:  9,  Y:  0,  Z:  0
X:  9,  Y:  0,  Z:  0
X:  9,  Y:  0,  Z:  0
X:  9,  Y:  0,  Z:  0
X:  9,  Y:  0,  Z:  0
X:  9,  Y:  0,  Z:  0
X:  9,  Y:  0,  Z:  0
X:  9,  Y:  0,  Z:  0
X:  9,  Y:  0,  Z:  0
X:  9,  Y:  0,  Z:  0
X:  9,  Y:  0,  Z:  0
X:  9,  Y:  0,  Z:  0
X:  9,  Y:  0,  Z:  0
X:  9,  Y:  0,  Z:  0
X:  9,  Y:  0,  Z:  0
```

Figure 6.12: Accelerometer output values for left gesture

15. If you place the phone as shown in *Figure 6.13* (opposite of the left gesture, which is the right gesture), the accelerometer will produce negative values on the *x*-axis:

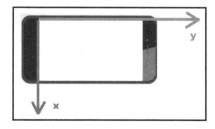

Figure 6.13: Right gesture (created by Freepik)

16. *Figure 6.14* shows the output values on each axis for the right gesture:

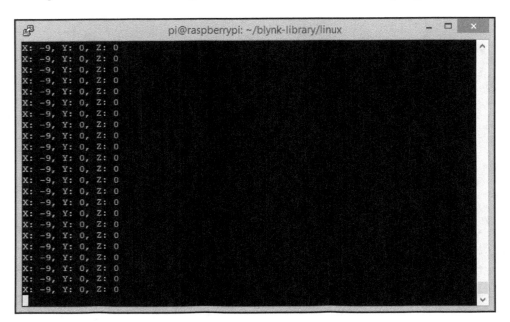

Figure 6.14: Output values for the right gesture

Calculating overall acceleration

Acceleration is a vector quantity, so it has a magnitude (size, length) and a direction. To get the overall magnitude, irrespective of orientation, we need to do a simple calculation.

If you want to get the overall acceleration with only x and y axes (known as 2D acceleration), you can calculate the magnitude (length) of the result with Pythagoras' rule:

$$acceleration = \sqrt{x^2 + y^2}$$

Figure 6.15 represents 2D acceleration using Pythagoras' rule:

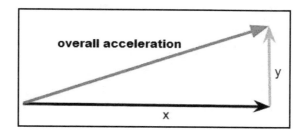

Figure 6.15: Representing 2D acceleration using Pythagoras' rule

If you want to get the overall acceleration using three axes (known as 3D acceleration), the magnitude can be calculated as:

$$acceleration = \sqrt{x^2 + y^2 + z^2}$$

Listing 6.2 shows sample code that can be run on Raspberry Pi to calculate the 2D and 3D magnitude:

Listing 6.2: Calculating 2D and 3D magnitude

Light sensor

Smartphones and tablets have built-in light sensors (also known as ambient light sensors) to measure the light level and adjust the brightness of the screen accordingly to save the battery. Blynk provides a Light widget to take values from the built-in ambient light sensor on your smartphone. The light level can be measured in lux (symbol: lx).

Adding the Light Sensor widget

The following steps will show you how to add a Light widget to your Blynk app:

1. In the Widget Box, under **SMARTPHONE SENSORS**, tap **Light Sensor**
2. The Light Sensor widget will be added to the canvas

Configuring the Light Sensor widget

The Light Sensor widget only works with virtual pins:

1. Tap the Light Sensor widget to open the **Light Sensor Settings** page.
2. Under **INPUT**, tap **PIN** and in the **Select pin list**, select **Virtual** followed by **V1**.
3. Tap **OK**. Your **Light Sensor Settings** page should look like this (*Figure 6.16*):

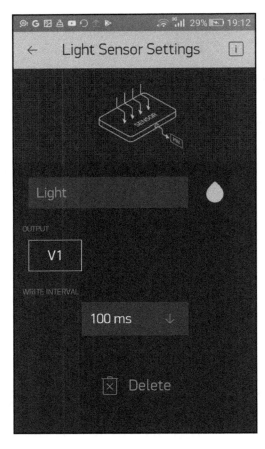

Figure 6.16: Configured Light Settings page

4. Tap the back arrow on the toolbar to go to the canvas. Now, your Light Sensor widget is configured with the virtual pin V1 as shown here (*Figure 6.17*):

Figure 6.17: Configured Light Sensor widget

Reading light sensor

You can read the values coming from the light sensor with virtual pins. It is very simple, as shown here. Normally, you will get the light level in lux:

```
BLYNK_WRITE(V1) {
//light value
int lx = param.asInt();
}
```

Listing 6.3 shows sample code that you can use to read the `light level` and display it on the console:

Listing 6.3: `Reading light level`

Then perform the following steps:

1. Now, save the file and exit the nano text editor by pressing *Ctrl + O*, followed by *Enter*, followed by *Ctrl +X*.

2. Build the project with the following command:

```
pi@raspberrypi:~/blynk-library/linux $ ./build.sh raspberry
```

3. Then, run the program with the auth token associated with your Blynk project:

```
pi@raspberrypi:~/blynk-library/linux $sudo ./blynk --
token=ca7bed1c92214503a65de8e20164994f
```

4. Tap the play button on the Blynk app builder to start the Blynk app on your mobile device. The console will output the current light level in lux (*Figure 6.18*). Cover the light sensor on your smartphone with your finger. You should get a value close to 0 for the light level:

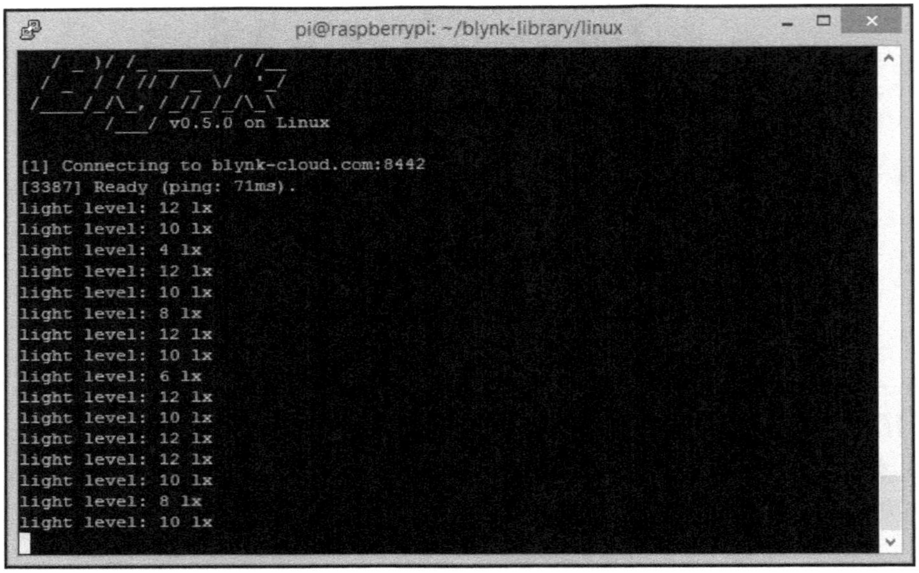

Figure 6.18: Reading light levels

Proximity sensor

Proximity sensors are also known as position sensors and allow you to determine how close your smartphone is to an object (such as your face). The distance can be measured in cm. However, most proximity sensors only return **Far/Near** information. Therefore, the return value will be 0 or 1, where 0/LOW is **Far** and 1/HIGH is **Near**.

As usual, you can add a Proximity Sensor widget to your project canvas from the Widget Box (*Figure 6.19*):

Figure 6.19: Proximity Sensor widget

Configuring the Proximity Sensor widget

The following steps will show you how to configure the Proximity Sensor widget with a virtual pin:

1. Tap the Proximity Sensor widget to open the **Proximity Sensor Settings** page.
2. Under **INPUT**, tap **PIN** and from the list view, select **Virtual** followed by **V1**. Keep the **WRITE INTERVAL** as **100 ms**.
3. Then, tap **OK**. Your **Proximity Sensor Settings** page should look like this (*Figure 6.20*):

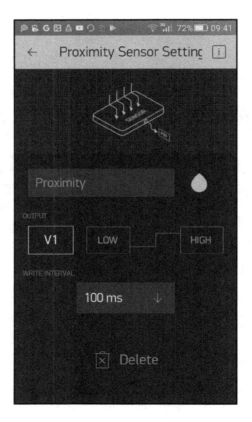

Figure 6.20: Virtual pin configuration for proximity sensor

4. Tap the back arrow on the toolbar to go to the canvas.

Writing code

Proximity can be easily determined with the virtual pins. The code snippet shown here can be used to take and process the incoming values to get the distance (near or far).

You can modify the `main.cpp` file in your Blynk project with the `BLYNK_WRITE(V1)` function:

```
BLYNK_WRITE(V1) {

// distance to object

int proximity = param.asInt();

if (proximity) {

print("there is some object nearby")

//NEAR

} else {

//FAR

print("there is no object nearby")

}

}
```

Then, build and run the code with the auth token associated with your project.

Summary

In this chapter, you learned how to take values from some of the built-in sensors on your smartphone or tablet. In `Chapter 7`, *Setting Up a Personal Blynk Server*, you will learn how to set up a personal Blynk server with Raspberry Pi without connecting to the Blynk cloud.

7
Setting Up a Personal Blynk Server

This chapter shows you how to install and configure a personal Blynk server on Raspberry Pi.

With this architecture, you can directly connect your hardware (that is, your Raspberry Pi) to the machine running the Blynk server through a network (whether an intranet or the internet) without using a third-party Blynk cloud.

In this chapter, you will learn about the following:

- Installing Java on Raspberry Pi
- Installing a Blynk personal server on Raspberry Pi
- Configuring the server
- Creating a Blynk app to connect with a personal Blynk server

You can run the Blynk personal server on Windows and Mac operating systems. The installation instructions can be found at `https://github.com/blynkkk/blynk-server#blynk-server`.

Setting up a Blynk server on Raspberry Pi

The following steps will show you how to set up a Blynk server on a Raspberry Pi. The Blynk server supports all versions of Raspberry Pi boards. However, Raspberry Pi 3 provides better performance for running dedicated servers than previous boards.

For testing purposes, you can use the same Raspberry Pi board to run both the server and the firmware for the hardware:

1. Using PuTTY, log in to your Raspberry Pi through SSH.
2. Install Java 8 using the following command:

   ```
   pi@raspberrypi:~ $sudo apt-get install oracle-java8-jdk
   ```

3. Verify the Java version using the following command:

   ```
   pi@raspberrypi:~ $ java -version
   ```

4. You will get the following output:

   ```
   java version "1.8.0_65"
   Java(TM) SE Runtime Environment (build 1.8.0_65-b17)
   Java HotSpot(TM) Client VM (build 25.65-b01, mixed mode)
   ```

5. Download the Blynk server JAR file using the following command:

   ```
   pi@raspberrypi:~ $wget
   "https://github.com/blynkkk/blynk-server/releases/download/v0.3
   3.4/server-0.33.4-java8.jar"
   ```

6. A file named `server-0.33.4-java8.jar` for the Blynk server will be downloaded to the Raspberry Pi.
7. Run the server (JAR file) on default hardware port `8080` and default application port `9443` (SSL port):

   ```
   pi@raspberrypi:~ $java -jar server-0.33.4-java8.jar -dataFolder
   /home/pi/Blynk
   ```

8. You will get following output for the first time:

   ```
   Blynk Server 0.34.0-SNAPSHOT successfully started.
   All server output is stored in folder '/home/pi/logs' file.
   Your Admin login email is admin@blynk.cc
   Your Admin password is admin
   ```

9. The login email and password for the admin is used to log in to the Blynk app builder running on your smartphone.
10. For subsequent restarts, you will get the following output:

    ```
    Blynk Server 0.34.0-SNAPSHOT successfully started.
    All server output is stored in folder '/home/pi/logs' file.
    ```

Enabling autostart with rc.local

You can configure the Raspberry Pi to autostart your Blynk server when the system starts:

1. On your Raspberry Pi, edit the /etc/rc.local file using the nano editor:

   pi@raspberrypi:~ $sudonano /etc/rc.local

2. Add the following command after the comment, but leave the exit 0 line at the end, then save (*Ctrl + O*, followed by *Enter*) the file and *Exit* (*Ctrl + X*):

   java -jar /home/pi/server-0.33.4.jar -dataFolder /home/pi/Blynk &

Editing the rc.local file

3. Enter the following to reboot the Pi:

```
pi@raspberrypi:~ $sudo reboot
```

Enabling autostart with crontab

If the Blynk server doesn't start with the `rc.local` configuration, you can use the `crontab`. The `crontab` can be used to schedule commands or scripts to run periodically and at fixed intervals:

1. Enter the following to open the `crontab`:

```
pi@raspberrypi:~ $ crontab -e
```

2. Then, select an editor by entering the option number. Enter the number 2 to open the `crontab` with the nano editor:

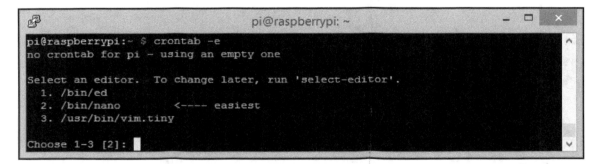

Choosing an editor to open the crontab

3. Add the following at the end of the file:

```
@reboot java -jar /home/pi/server-0.33.4.jar -dataFolder
/home/pi/Blynk &
```

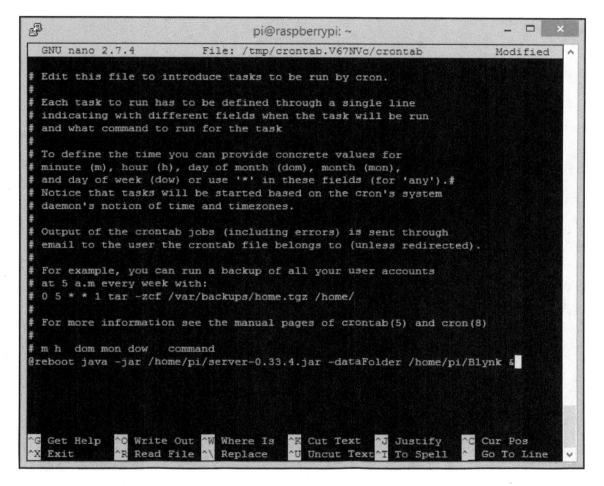

Editing crontab

4. Press *Ctrl + O*, followed by *Enter*, followed by *Ctrl + X*.
5. Enter the following to reboot the Pi:

```
pi@raspberrypi:~ $sudo reboot
```

Verifying that the Blynk server is running

When you restart the Raspberry Pi by configuring it using one of the previous two methods, the Blynk server will be autostarted. You can verify this by running the following command:

```
pi@raspberrypi:~ $ps -aux | grep java
```

You should see the following entry for the Blynk server:

```
pi 843 34.2 6.1 358340 58644 pts/0 Sl+ 17:05 0:23 java -jar server-0.33.4-
java8.jar -dataFolder /home/pi/Blynk
```

Reporting the process's status

Connecting the Blynk app builder with the server

Normally, the Blynk app builder works with the Blynk cloud by connecting to it via the internet. If you want to connect your Blynk app builder with your personal Blynk server, you must configure some attributes in the Blynk app builder. These attributes will help you to find the personal Blynk server that is deployed on the same network (that is, your home Wi-Fi network). If you want to access the personal Blynk server from the internet, you can use port forwarding to access it. To do this, you will also need a static IP address:

1. First, you should configure the custom server path for your personal Blynk server running on Raspberry Pi.
2. On the login page, tap **Create New Account**:

Creating a new account

3. Tap the three dots icon to view the server settings:

Server settings icon

4. Slide the toggle button from **BLYNK** to **CUSTOM**:

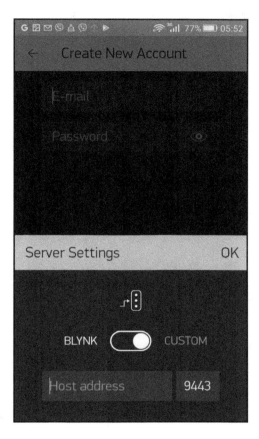

Server settings—step 1

5. Enter the IP address of your Raspberry Pi (that is, 192.168.1.4) and the port number. The default port for your Blynk server is 9443. Then, tap **OK** to save the settings:

Server settings—step 2

6. Type in a username (email address) and a password for the new user account.
7. Tap **Sign Up** to create the account. You can see the message **Connecting...** on the bottom of the page while creating the account.

8. Once you have created an account, the app will show you a tip about how energy works. Tap the **Cool! Got it** button and you will go to the project dashboard:

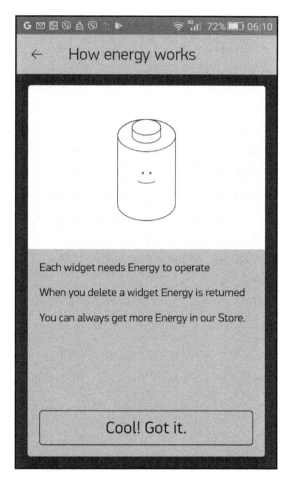

Tip about how energy works

Creating a new project to get the auth token

To run a project, you will need the auth token associated with a Blynk app builder project under your Blynk personal server:

1. Tap **New Project**. You'll get the **Create New Project** page:

Creating a New Project

2. In the **Create New Project** page, type in the **Project Name** and choose the hardware. In this case, you should select your Raspberry Pi model. Then, tap **OK**:

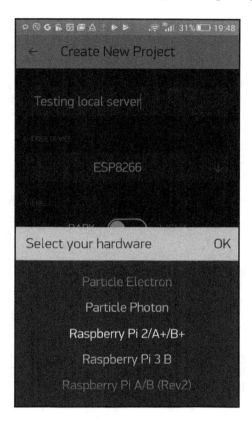

Project name and hardware model

3. Now, your **Create New Project** page should look like the preceding screenshot. Tap the **Create** button to create the project workspace:

Creating a New Project settings

4. You will get a notification about the auth token. Tap **OK** to close the message:

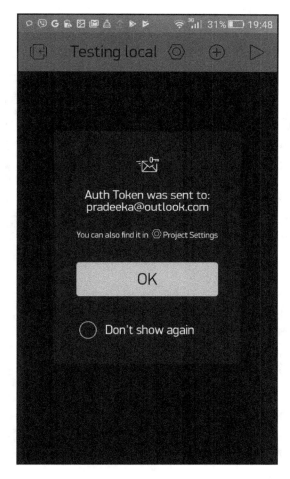

Notification message about the auth token

5. Tap the **Project Settings** (represented by a nut) icon on the toolbar to open the **Project Settings** page. Under **AUTH TOKENS**, tap **Copy all**. The auth token will be copied to the clipboard:

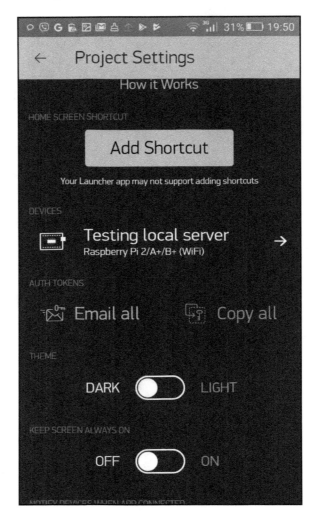

Project Settings

6. Open the Notepad app (or any text editor) and paste the copied item to view the auth token:

Auth token

Using the administration interface

Refer to the following steps:

1. You can open the administration interface by entering the following address in a web browser:

   ```
   https://IP_ADDRESS:9443/admin
   ```

2. Type in the default administrator credentials to log in. Here are the default login credentials for admin:
   - **Email address**: admin@blynk.cc
   - **Password**: admin

3. Click **Sign-in** to log in to the administrator interface:

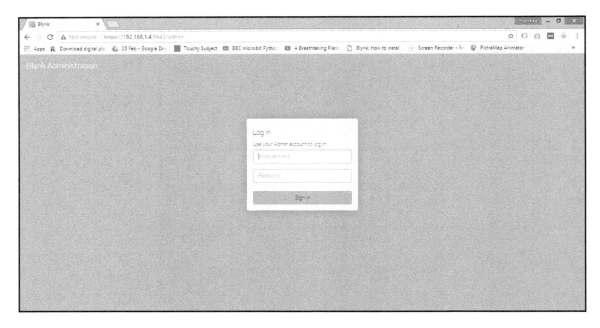

Administrator interface

4. You will then see the dashboard:

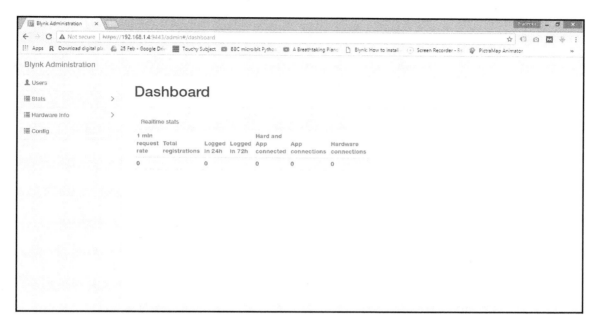

Dashboard

5. In the left-hand list, click **Users**. You will get the list of user accounts registered with your local Blynk server, including the admin:

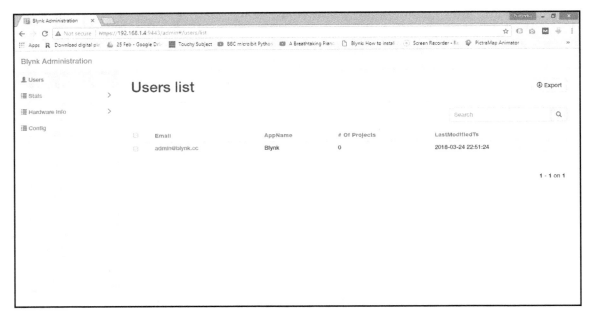

Users list

6. In the left-hand navigation pane, click **Config**. You will get a list of configurations for the following:
    - `twitter4j.properties`: This is required for Twitter notification
    - `single_token_mail_body.txt`
    - `server.properties`: Settings related to the server
    - `mail.properties`: Required for email notification
    - `gcm.properties`: Required for phone notification

- `db.properties`: A setting related to the database

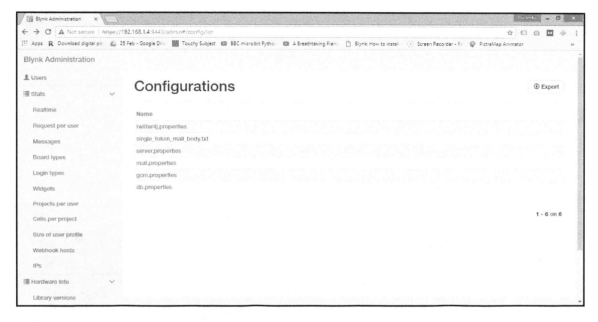

Configurations

Writing a simple code to build the connection

To connect the Raspberry Pi hardware to the server running on the same or a different Raspberry Pi, you should have the minimum code to make the connection. To set up the connection, go through the following steps:

1. Open the `main.cpp` file and remove all the existing lines. Then, type in the code under *Listing 7.1.* Use the following commands to open the source file:

```
pi@raspberrypi:~/blynk-library/linux $ cd blynk-library/linux
pi@raspberrypi:~/blynk-library/linux $sudonano main.cpp
```

Listing 7.1: Connect with local Blynk server
https://github.com/PacktPublishing/Hands-On-Internet-of-Things-with-Blynk/compare/master...TrushaShriyan:patch-1

2. Make sure that you correctly add the IP address with the following line:

   `Blynk.begin(auth, IPaddress(192.168.1.4), 8080);`

3. Press *Ctrl + O,* followed by *Enter,* followed by *Ctrl + X.*
4. Build the project with the following command:

   `pi@raspberrypi:~/blynk-library/linux $ ./build.sh raspberry`

5. Run the project with the auth token associated with your Blynk app project:

   `pi@raspberrypi:~/blynk-library/linux $sudo ./blynk --token=e6d7cfa77b9e4dce9ca15bd7bfdf27e6`

6. You will get the following output on the console:

Blynk application is up and running

Summary

In this chapter, you learned how to install and set up a personal Blynk server on your Raspberry Pi. Then, you built a simple application that can be used to connect to the Blynk personal server through a Wi-Fi network. In Chapter 8, *Controlling a Robot with Blynk,* you will learn how to build a simple robot vehicle and control it with a Blynk application.

Controlling a Robot with Blynk

8

The Blynk architecture provides an easy way to build robotic applications by combining hardware and software. As a beginner's guide, this chapter shows you how to build a simple robot vehicle with Raspberry Pi hardware. The control interface can be easily built with the Blynk app builder and you can control the vehicle through a Wi-Fi network. In this chapter, you will learn:

- How to choose a chassis kit
- How to create a control application using the Blynk app builder
- How to use a H-bridge motor driver to drive two DC motors
- How to control motors with a C++ application

Choosing a chassis kit

Robot chassis kits provide an easy way to build the body of your vehicle. Most chassis kits come with DC gear motors and wheels. Some provide built-in battery compartments. To build a simple vehicle, you can use a two-wheeled robot chassis kit.

Adafruit, SparkFun, and Pololu sell great two-wheeled chassis kits that you can purchase for this project.

Adafruit

This includes the following kits:

- **Mini Round Robot Chassis Kit - 2WD with DC Motors** (https://www.adafruit.com/product/3216):

  This chassis kit provides you everything you need to build the two-wheeled robot vehicle. The chassis can be built by combining metal plates with standoffs. The kit includes the following things:

    - 2 x drive motors (drive with 3-6 VDC, 200-400 mA run, 1.5 A hard stall)
    - 2 x wheels
    - 1 x plastic caster ball
    - Anodized aluminum frame and all mounting hardware for assembly

- **Mini 3-Layer Round Robot Chassis Kit - 2WD with DC Motors** (https://www.adafruit.com/product/3244):

  This is the extended version of the mini round robot chassis kit and provides more vertical space to hold your electronics and battery compartments. This kit includes:

    - 2 x drive motors (drive with 3-6VDC, 200-400 mA run, 1.5A hard stall)
    - 2 x wheels
    - 1 x plastic caster ball
    - Anodized aluminum frames and all mounting hardware for assembly

- **Mini Robot Rover Chassis Kit - 2WD with DC Motors** (https://www.adafruit.com/product/2939):

The mini robot rover chassis kit provides you everything you need to build a two-wheel-drive robot rover. The chassis can be vertically expanded by adding more metal plates (https://www.adafruit.com/product/2944). This kit includes the following things:

- 2 x wheels
- 2 x DC motors in micro servo shape
- 1 x support wheel
- 1 x metal chassis
- 1 x top metal plate with mounting hardware

SparkFun

This includes the following kits:

- **Circular Robotics Chassis Kit (Two-Layer)** (https://www.sparkfun.com/products/14332):

  This kit is identical to Adafruit's mini round robot chassis kit but it doesn't include the DC motors. You can use a pair of Hobby Gearmotors (https://www.sparkfun.com/products/13302) with this kit:

Image courtesy of SparkFun Electronics: https://creativecommons.org/licenses/by/2.0/

- **Circular Robotics Chassis Kit (Three-Layer)** (https://www.sparkfun.com/products/14339):

  This chassis kit provides an additional chassis layer to vertically expand the space of the vehicle. You can use this space to attach additional electronics or battery packs:

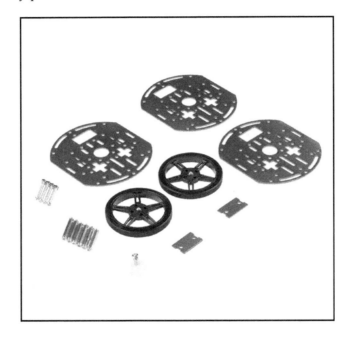

  This kit doesn't include DC gearmotors. You can purchase a pair of motors at https://www.sparkfun.com/products/13302.

- **Shadow Chassis** (https://www.sparkfun.com/products/13301):

  The shadow chassis kit is built with ABS plastic and you can simply snap each panel together to build the shell of the vehicle. This kit doesn't provide any DC gearmotors or wheels:

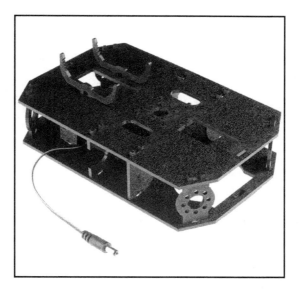

Image courtesy of SparkFun Electronics: https://creativecommons.org/licenses/by/2.0/

The kit includes the following things:

- 1 x top chassis plate
- 1 x bottom chassis plate
- 2 x front motor mount
- 2 x rear motor mount
- 4 x side strut
- 2 x encoder mount
- 2 x micro-controller mount
- 1 x battery pack clip
- 1 x line follower mount
- 1 x line follower mount plate
- 1 x nub caster
- 1 x 4xAA battery holder

Pololu

Romi Chassis Kits (`https://www.pololu.com/category/203/romi-chassis-kits`):

- The Romi Chassis kit includes all the basic mechanical parts to build a vehicle shell, including two motors, two wheels, one ball caster, and battery contacts
- You can choose one of the chassis kits listed here to build the vehicle will be discussing in the coming sections

Creating a Blynk app

The control user interface can be created with the Blynk app builder using four button controllers to move the robot forward, backward, left, and right:

1. Open the Blynk app builder and tap **New Project**:

Choosing New Project

2. In the **Create New Project** page, type in the project name, `Robot Car`. Under **CHOOSE DEVICE**, select the correct Raspberry Pi model. Then, under **CONNECTION TYPE**, select **WiFi**. If you're using a Raspberry Pi model 1 or 2, you should connect an external Wi-Fi dongle to the Raspberry Pi's USB port. The Raspberry Pi model 3 has a built-in Wi-Fi module, so you don't need to connect a Wi-Fi dongle to it. Finally, tap **Create** to build the project workspace:

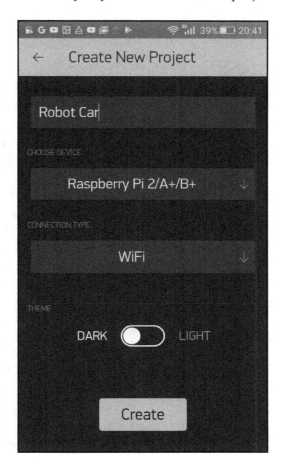

Project Settings page

3. Tap the **OK** button:

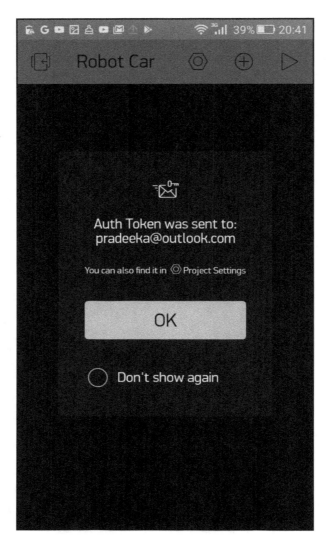

Auth token sending notification

4. In the project workspace, tap the plus icon to open the Widget Box and add four buttons:

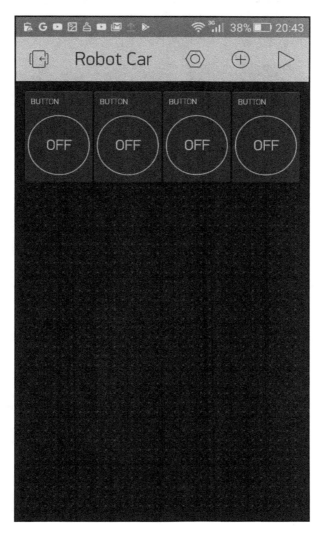

Inserting four buttons

5. Then, arrange the buttons as shown in the following screenshot:

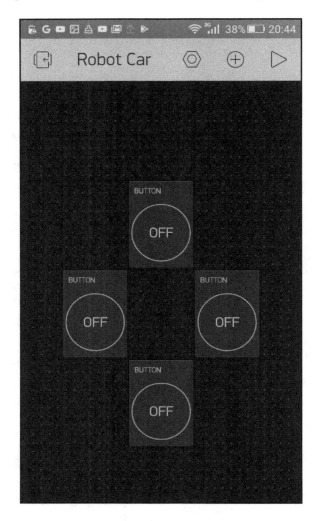

Arranging four buttons to make the controller

- Configure the buttons as follows:
    - **FORWARD**: Virtual pin V0
    - **BACKWARD**: Virtual pin V3
    - **LEFT**: Virtual pin V1
    - **RIGHT**: Virtual pin V2

6. Once configured, the buttons should look like this:

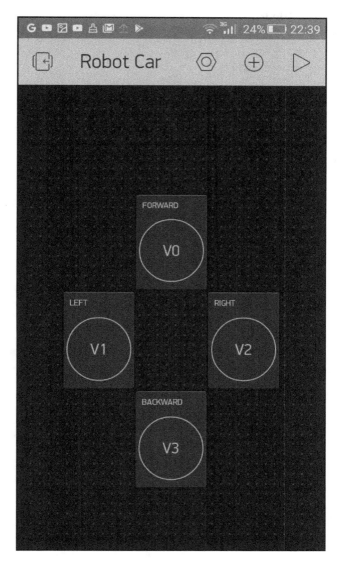

Buttons after configuration

7. Now, you have successfully built the controller interface. The left and right buttons can be used to turn the robot forward left and forward right. You will use the swing turn mechanism to turn the robot to either direction:

- **Forward left swing turn**: Turn off the left motor, drive the right motor forward:

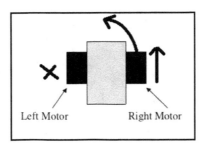

Forward left swing turn

- **Forward right swing turn**: Turn off the right motor, drive the left motor forward:

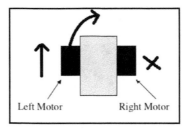

Forward right swing turn

Motor driver

The TB6612FNG Dual Motor Driver Carrier (https://www.pololu.com/product/713) allows you to drive a pair of small brushed DC motors. This type of motor driver is known as a full H-bridge motor driver. This driver can accept input voltage ranges from 4.5V to 13.5V to drive DC motors, and the logic supply voltage can be 2V to 7V.

A 1×16-pin breakaway 0.1″ male header strip is included with the TB6612FNG motor driver carrier. You can find more details about the breakaway male header at `https://www.pololu.com/product/965`. This strip can be soldered to the carrier board so that it can be used with solderless breadboards.

Once soldered, you can use it with a solderless breadboard. You can choose a breadboard to fit the size and shape of your robot chassis. Some breadboards offer mounting options, so you can easily mount them using standoffs. Pololu offers a wide array of breadboards (`https://www.pololu.com/category/28/solderless-breadboards`).

The following diagram shows the top view of the board with pads labeled:

Top view with pads labeled

The following table shows the functions of each connection pad:

Pin	I/O	Function
AO1	O	Connect with left motor
AO2		
BO1	O	Connect with right motor
BO2		
PWMA	I	Connect with Raspberry Pi
AIN2		
AIN1		

BIN1		
BIN2	I	Connect with Raspberry Pi
PWMB		
VCC	-	Small signal supply connect with Raspberry Pi 5V pin
VMOT	-	Power supply for motors. 2.5V to 13.5V DC from a battery pack.
GND	-	Power ground

Motors should be directly connected with the motor driver. The connection between the motor driver and the Raspberry Pi handles the control signal:

- Connecting motors:
    - Connect the wires coming from the left motor to AO1 and AO2
    - Connect the wires coming from the right motor to BO1 and BO2
- Connecting battery pack
    - Connect the battery pack to VMOT and GND
- Connecting Raspberry Pi
    - Connect VCC to the Raspberry Pi 5V
    - To control the left motor, connect:
        - AIN1 to GPIO 4
        - AIN2 to GPIO 5
        - PWMA to GPIO 6
    - To control the right motor, connect:
        - BIN1 to GPIO 27
        - BIN2 to GPIO 28
        - PWMB to GPIO 29
    - Connect STBY to the Raspberry Pi 5V

The following screenshot shows the pinout of the Raspberry Pi GPIO header. WiringPi and BCM use different pin number assignments. The C++ code you will be writing uses WiringPi notation for the pin numbering:

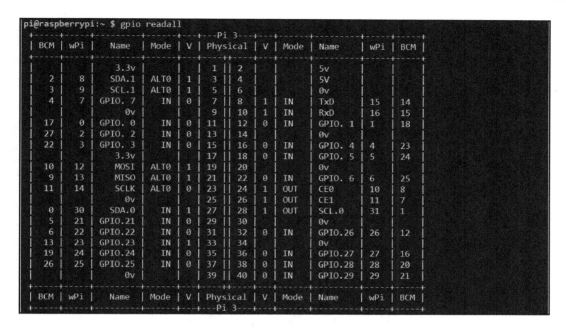

```
pi@raspberrypi:~ $ gpio readall
+-----+-----+---------+------+---+---Pi 3---+---+------+---------+-----+-----+
| BCM | wPi |   Name  | Mode | V | Physical | V | Mode |   Name  | wPi | BCM |
+-----+-----+---------+------+---+----++----+---+------+---------+-----+-----+
|     |     |    3.3v |      |   |  1 || 2  |   |      | 5v      |     |     |
|   2 |   8 |   SDA.1 | ALT0 | 1 |  3 || 4  |   |      | 5V      |     |     |
|   3 |   9 |   SCL.1 | ALT0 | 1 |  5 || 6  |   |      | 0v      |     |     |
|   4 |   7 |  GPIO. 7 |  IN | 0 |  7 || 8  | 1 | IN   | TxD     | 15  | 14  |
|     |     |      0v |      |   |  9 || 10 | 1 | IN   | RxD     | 16  | 15  |
|  17 |   0 |  GPIO. 0 |  IN | 0 | 11 || 12 | 0 | IN   | GPIO. 1 | 1   | 18  |
|  27 |   2 |  GPIO. 2 |  IN | 0 | 13 || 14 |   |      | 0v      |     |     |
|  22 |   3 |  GPIO. 3 |  IN | 0 | 15 || 16 | 0 | IN   | GPIO. 4 | 4   | 23  |
|     |     |    3.3v |      |   | 17 || 18 | 0 | IN   | GPIO. 5 | 5   | 24  |
|  10 |  12 |    MOSI | ALT0 | 1 | 19 || 20 |   |      | 0v      |     |     |
|   9 |  13 |    MISO | ALT0 | 1 | 21 || 22 | 0 | IN   | GPIO. 6 | 6   | 25  |
|  11 |  14 |    SCLK | ALT0 | 0 | 23 || 24 | 1 | OUT  | CE0     | 10  | 8   |
|     |     |      0v |      |   | 25 || 26 | 1 | OUT  | CE1     | 11  | 7   |
|   0 |  30 |   SDA.0 |   IN | 1 | 27 || 28 | 1 | OUT  | SCL.0   | 31  | 1   |
|   5 |  21 |  GPIO.21 |  IN | 0 | 29 || 30 |   |      | 0v      |     |     |
|   6 |  22 |  GPIO.22 |  IN | 0 | 31 || 32 | 0 | IN   | GPIO.26 | 26  | 12  |
|  13 |  23 |  GPIO.23 |  IN | 1 | 33 || 34 |   |      | 0v      |     |     |
|  19 |  24 |  GPIO.24 |  IN | 0 | 35 || 36 | 0 | IN   | GPIO.27 | 27  | 16  |
|  26 |  25 |  GPIO.25 |  IN | 0 | 37 || 38 | 0 | IN   | GPIO.28 | 28  | 20  |
|     |     |      0v |      |   | 39 || 40 | 0 | IN   | GPIO.29 | 29  | 21  |
+-----+-----+---------+------+---+----++----+---+------+---------+-----+-----+
| BCM | wPi |   Name  | Mode | V | Physical | V | Mode |   Name  | wPi | BCM |
+-----+-----+---------+------+---+---Pi 3---+---+------+---------+-----+-----+
```

Raspberry Pi GPIO header

The following table shows the control signal for each mode of two motors. You can use **Clock Wise (CW)**, **Counter Clock Wise (CCW)**, and stop modes to control the motors:

Input				Output		
IN1	IN2	PWM	STBY	OUT1	OUT2	Mode
H	H	H/L	H	L	L	Short brake
L	H	H	H	L	H	CCW
		L	H	L	L	Short brake
H	L	H	H	H	L	CW
		L	H	L	L	Short brake
L	L	H	H	OFF (High impedance)		Stop
H/L	H/L	H/L	L	OFF (High impedance)		Standby

Control signal for two motors

Listing 8.1 (`https://github.com/PacktPublishing/Hands-On-Internet-of-Things-with-Blynk/blob/master/Chapter%208/Listing%208-1/main.cpp`) shows the C++ code that you can use to run on the Raspberry Pi to control the two motors according to the signal coming from the Blynk app:

Listing 8.1: `Robot vehicle control code`

Refer to the following steps:

1. Open `main.cpp` and replace the existing code with the *Listing 8.1*. Then, build the project again.
2. Finally, run the project with the auth token.
3. Tap the play button in the Blynk app builder to run the controller application:

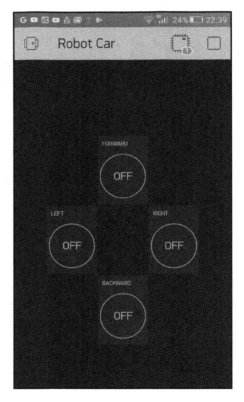

Controller app in play mode

4. Tap each button to move the robot vehicle.

Summary

In this chapter, you learned how to build and control a robot vehicle with a Blynk application. Throughout this book, you have used various Blynk controllers to build rich user interfaces to control hardware and display sensor data. You also used digital and virtual pins to communicate with controllers.

The Blynk platform is rapidly growing and you can always check out the latest Blynk technologies by visiting their official website (`https://www.blynk.cc/`) and support forum (`https://community.blynk.cc/`).

Other Books You May Enjoy

If you enjoyed this book, you may be interested in these other books by Packt:

Internet of Things for Architects
Perry Lea

ISBN: 978-1-78847-059-9

- Understand the role and scope of architecting a successful IoT deployment, from sensors to the cloud
- Scan the landscape of IoT technologies that span everything from sensors to the cloud and everything in between
- See the trade-offs in choices of protocols and communications in IoT deployments
- Build a repertoire of skills and the vernacular necessary to work in the IoT space
- Broaden your skills in multiple engineering domains necessary for the IoT architect

Mastering Internet of Things
Peter Waher

ISBN: 978-1-78839-748-3

- Create your own project, run and debug it
- Master different communication patterns using the MQTT, HTTP, CoAP, LWM2M and XMPP protocols
- Build trust-based as hoc networks for open, secure and interoperable communication
- Explore the IoT Service Platform
- Manage the entire product life cycle of devices
- Understand and set up the security and privacy features required for your system
- Master interoperability, and how it is solved in the realms of HTTP,CoAP, LWM2M and XMPP

Leave a review - let other readers know what you think

Please share your thoughts on this book with others by leaving a review on the site that you bought it from. If you purchased the book from Amazon, please leave us an honest review on this book's Amazon page. This is vital so that other potential readers can see and use your unbiased opinion to make purchasing decisions, we can understand what our customers think about our products, and our authors can see your feedback on the title that they have worked with Packt to create. It will only take a few minutes of your time, but is valuable to other potential customers, our authors, and Packt. Thank you!

Index

www.ingramcontent.com/pod-product-compliance
Lightning Source LLC
LaVergne TN
LVHW081520050326
832903LV00025B/1560